STUDENT SOLUTIONS MANUAL

FOR DEVORE'S

PROBABILITY
AND
STATISTICS

for Engineering and the Sciences

Sixth Edition

Julie Ann Seely

THOMSON

™

BROOKS/COLE

Australia • Canada • Mexico • Singapore • Spain • United Kingdom • United States

Printed in Canada
2 3 4 5 6 7   07 06 05 04

Printer: Transcontinental Printing/Louiseville

ISBN: 0-534-39934-7

For more information about our products,
contact us at:
Thomson Learning Academic Resource Center
1-800-423-0563

For permission to use material from this text,
contact us by:
Phone: 1-800-730-2214
Fax: 1-800-730-2215
Web: http://www.thomsonrights.com

Brooks/Cole—Thomson Learning
10 Davis Drive
Belmont, CA 94002-3098
USA

Asia
Thomson Learning
5 Shenton Way #01-01
UIC Building
Singapore 068808

Australia/New Zealand
Thomson Learning
102 Dodds Street
Southbank, Victoria 3006
Australia

Canada
Nelson
1120 Birchmount Road
Toronto, Ontario M1K 5G4
Canada

Europe/Middle East/South Africa
Thomson Learning
High Holborn House
50/51 Bedford Row
London WC1R 4LR
United Kingdom

Latin America
Thomson Learning
Seneca, 53
Colonia Polanco
11560 Mexico D.F.
Mexico

Spain/Portugal
Paraninfo
Calle/Magallanes, 25
28015 Madrid, Spain

CONTENTS

CHAPTER 1

Section 1.1

1.

    a. Houston Chronicle, Des Moines Register, Chicago Tribune, Washington Post

    b. Capital One, Campbell Soup, Merrill Lynch, Pulitzer

    c. Bill Jasper, Kay Reinke, Helen Ford, David Menedez

    d. 1.78, 2.44, 3.5, 3.04

3.

    a. In a sample of 100 VCRs, what are the chances that more than 20 need service while under warrantee? What are the chances than none need service while still under warrantee?

    b. What proportion of all VCRs of this brand and model will need service within the warrantee period?

5.

    a. No, the relevant conceptual population is all scores of all students who participate in the SI in conjunction with this particular statistics course.

    b. The advantage to randomly choosing students to participate in the two groups is that we are more likely to get a sample representative of the population at large. If it were left to students to choose, there may be a division of abilities in the two groups which could unnecessarily affect the outcome of the experiment.

    c. If all students were put in the treatment group there would be no results with which to compare the treatments.

7.   One could generate a simple random sample of all single family homes in the city or a stratified random sample by taking a simple random sample from each of the 10 district neighborhoods. From each of the homes in the sample the necessary variables would be collected. This would be an enumerative study because there exists a finite, identifiable population of objects from which to sample.

9.

    a. There could be several explanations for the variability of the measurements. Among them could be measuring error, (due to mechanical or technical changes across measurements), recording error, differences in weather conditions at time of measurements, etc.

    b. This could be called an analytic study because there is no sampling frame.

Section 1.2

11.

6l	034
6h	667899
7l	00122244
7h	
8l	001111122344
8h	5557899
9l	03
9h	58

Stem=Tens
Leaf=Ones

This display brings out the gap in the data: There are no scores in the high 70's.

13.

a.

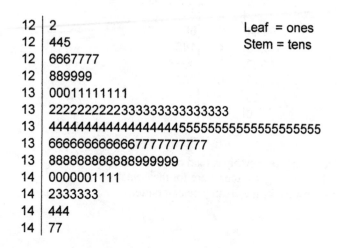

```
12 | 2                                    Leaf = ones
12 | 445                                  Stem = tens
12 | 6667777
12 | 889999
13 | 00011111111
13 | 2222222222333333333333333
13 | 444444444444444445555555555555555555
13 | 66666666666667777777777
13 | 888888888888999999
14 | 0000001111
14 | 2333333
14 | 444
14 | 77
```

The observations are highly concentrated at 134 – 135, where the display suggests the typical value falls.

b.

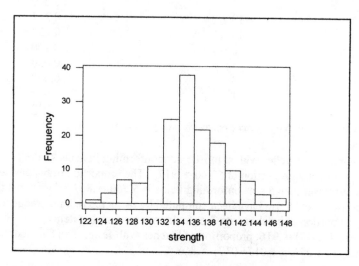

The histogram is symmetric and unimodal, with the point of symmetry at approximately 135.

15.

Crunchy		Creamy
	2	2
644	3	69
77220	4	145
6320	5	3666
222	6	258
55	7	
0	8	

Both sets of scores are reasonably spread out. There appear to be no outliers. The three highest scores are for the crunchy peanut butter, the three lowest for the creamy peanut butter.

17.

a.

Number Nonconforming	Frequency	RelativeFrequency(Freq/60)
0	7	0.117
1	12	0.200
2	13	0.217
3	14	0.233
4	6	0.100
5	3	0.050
6	3	0.050
7	1	0.017
8	1	0.017

doesn't add exactly to 1 because relative frequencies have been rounded 1.001

b. The number of batches with at most 5 nonconforming items is 7+12+13+14+6+3 = 55, which is a proportion of 55/60 = .917. The proportion of batches with (strictly) fewer than 5 nonconforming items is 52/60 = .867. Notice that these proportions could also have been computed by using the relative frequencies: e.g., proportion of batches with 5 or fewer nonconforming items = 1- (.05+.017+.017) = .916; proportion of batches with fewer than 5 nonconforming items = 1 - (.05+.05+.017+.017) = .866.

4

c. The following is a Minitab histogram of this data. The center of the histogram is somewhere around 2 or 3 and it shows that there is some positive skewness in the data. Using the rule of thumb in Exercise 1, the histogram also shows that there is a lot of spread/variation in this data.

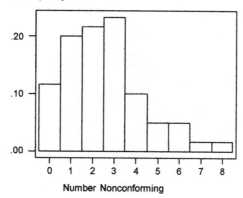

19.

a. From this frequency distribution, the proportion of wafers that contained at least one particle is (100-1)/100 = .99, or 99%. Note that it is much easier to subtract 1 (which is the number of wafers that contain 0 particles) from 100 than it would be to add all the frequencies for 1, 2, 3,... particles. In a similar fashion, the proportion containing at least 5 particles is (100 - 1-2-3-12-11)/100 = 71/100 = .71, or, 71%.

b. The proportion containing between 5 and 10 particles is (15+18+10+12+4+5)/100 = 64/100 = .64, or 64%. The proportion that contain strictly between 5 and 10 (meaning strictly *more* than 5 and strictly *less* than 10) is (18+10+12+4)/100 = 44/100 = .44, or 44%.

c. The following histogram was constructed using Minitab. The data was entered using the same technique mentioned in the answer to exercise 8(a). The histogram is *almost* symmetric and unimodal; however, it has a few relative maxima (i.e., modes) and has a very slight positive skew.

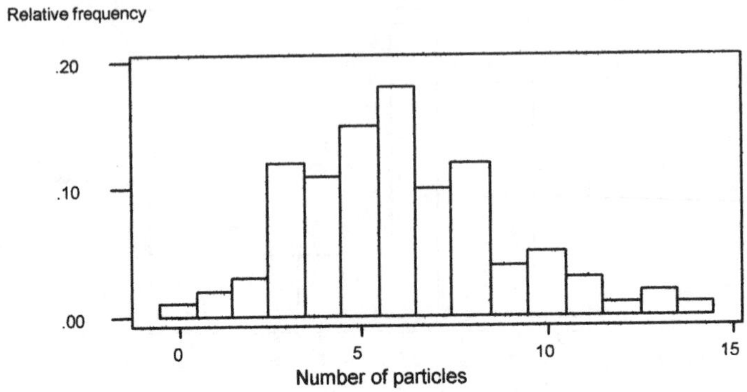

Relative frequency

Number of particles

21.

a. A histogram of the y data appears below. From this histogram, the number of subdivisions having no cul-de-sacs (i.e., $y = 0$) is $17/47 = .362$, or 36.2%. The proportion having at least one cul-de-sac ($y \geq 1$) is $(47-17)/47 = 30/47 = .638$, or 63.8%. Note that subtracting the number of cul-de-sacs with $y = 0$ from the total, 47, is an easy way to find the number of subdivisions with $y \geq 1$.

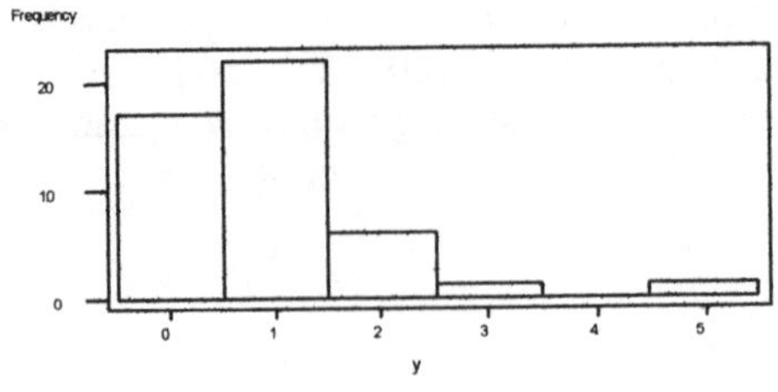

Frequency

y

6

b. A histogram of the z data appears below. From this histogram, the number of subdivisions with at most 5 intersections (i.e., $z \leq 5$) is 42/47 = .894, or 89.4%. The proportion having fewer than 5 intersections ($z < 5$) is 39/47 = .830, or 83.0%.

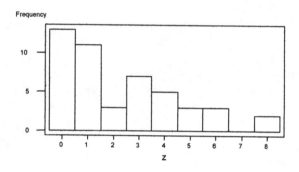

23.

a.

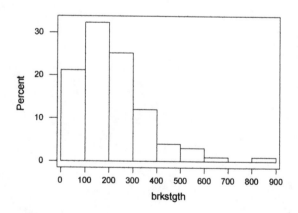

The histogram is skewed right, with most observations between 0 and 300 cycles. The class holding the most observations is between 100 and 200 cycles.

b.

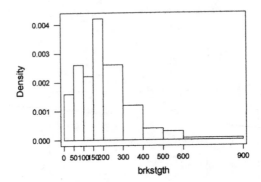

c   [proportion $\geq$ 100] = 1 – [proportion < 100] = 1 - .21 = .79

25.   Histogram of original data:

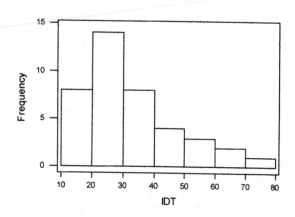

Histogram of transformed data:

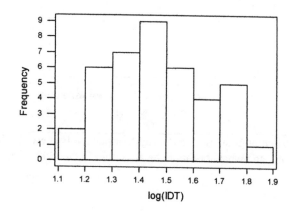

The transformation creates a much more symmetric, mound-shaped histogram.

27.

a. The endpoints of the class intervals overlap. For example, the value 50 falls in both of the intervals '0 – 50' and '50 – 100'.

b.

Class Interval	Frequency	Relative Frequency
0 - < 50	9	0.18
50 - < 100	19	0.38
100 - < 150	11	0.22
150 - < 200	4	0.08
200 - < 250	2	0.04
250 - < 300	2	0.04
300 - < 350	1	0.02
350 - < 400	1	0.02
>= 400	1	0.02
	50	1.00

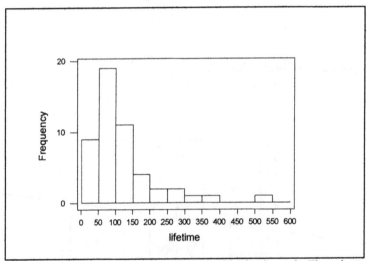

The distribution is skewed to the right, or positively skewed. There is a gap in the histogram, and what appears to be an outlier in the '500 – 550' interval.

c.

Class Interval	Frequency	Relative Frequency
2.25 - < 2.75	2	0.04
2.75 - < 3.25	2	0.04
3.25 - < 3.75	3	0.06
3.75 - < 4.25	8	0.16
4.25 - < 4.75	18	0.36
4.75 - < 5.25	10	0.20
5.25 - < 5.75	4	0.08
5.75 - < 6.25	3	0.06

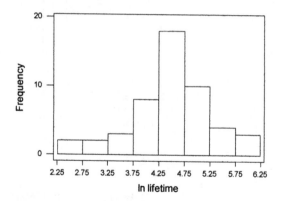

The distribution of the natural logs of the original data is much more symmetric than the original.

d. The proportion of lifetime observations in this sample that are less than 100 is .18 + .38 = .56, and the proportion that is at least 200 is .04 + .04 + .02 + .02 + .02 = .14.

29.

Complaint	Frequency	Relative Frequency
B	7	0.1167
C	3	0.0500
F	9	0.1500
J	10	0.1667
M	4	0.0667
N	6	0.1000
O	21	0.3500
	60	1.0000

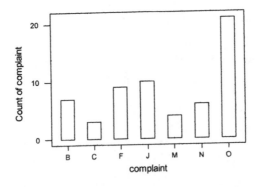

31.

Class	Frequency	Relative Frequency	Cumulative Relative Frequency
0.0 - under 4.0	2	2	0.050
4.0 - under 8.0	14	16	0.400
8.0 - under 12.0	11	27	0.675
12.0 - under 16.0	8	35	0.875
16.0 - under 20.0	4	39	0.975
20.0 - under 24.0	0	39	0.975
24.0 - under 28.0	1	40	1.000

Section 1.3

33.

    a. $\bar{x} = 192.57$, $\tilde{x} = 189$.   The mean is larger than the median, but they are still fairly close together.

    b. Changing the one value, $\bar{x} = 189.71$, $\tilde{x} = 189$.   The mean is lowered, the median stays the same.

    c. $\bar{x}_{tr} = 191.0$. $\frac{1}{14} = .07$ or 7% trimmed from each tail.

    d. For n = 13, $\Sigma x = (119.7692) \times 13 = 1{,}557$
         For n = 14, $\Sigma x = 1{,}557 + 159 = 1{,}716$

$$\bar{x} = \frac{1716}{14} = 122.5714 \text{ or } 122.6$$

35.

    a. The sample mean is $\bar{x} = (100.4/8) = 12.55$.

    The sample size (n = 8) is even. Therefore, the sample median is the average of the (n/2) and (n/2) + 1 values. By sorting the 8 values in order, from smallest to largest: 8.0 8.9 11.0 12.0 13.0 14.5 15.0 18.0, the forth and fifth values are 12 and 13. The sample median is (12.0 + 13.0)/2 = 12.5.

    The 12.5% trimmed mean requires that we first trim (.125)(n) or 1 value from the ends of the ordered data set. Then we average the remaining 6 values. The 12.5% trimmed mean $\bar{x}_{tr(12.5)}$ is 74.4/6 = 12.4.

    All three measures of center are similar, indicating little skewness to the data set.

    b. The smallest value (8.0) could be increased to any number below 12.0 (a change of less than 4.0) without affecting the value of the sample median.

    c. The values obtained in part (a) can be used directly. For example, the sample mean of 12.55 psi could be re-expressed as

$$(12.55 \text{ psi}) \times \left(\frac{1ksi}{2.2\,psi} \right) = 5.70ksi.$$

37. $\bar{x} = 12.01$, $\tilde{x} = 11.35$, $\bar{x}_{tr(10)} = 11.46$. The median or the trimmed mean would be good choices because of the outlier 21.9.

39.

a. $\Sigma x_i = 16.475$ so $\bar{x} = \dfrac{16.475}{16} = 1.0297$

$$\tilde{x} = \dfrac{(1.007 + 1.011)}{2} = 1.009$$

b. 1.394 can be decreased until it reaches 1.011(the largest of the 2 middle values) – i.e. by $1.394 - 1.011 = .383$, If it is decreased by more than .383, the median will change.

41.

a. $\dfrac{7}{10} = .70$

b. $\bar{x} = .70$ = proportion of successes

c. $\dfrac{s}{25} = .80$ so s = (0.80)(25) = 20

total of 20 successes

$20 - 7 = 13$ of the new cars would have to be successes

43. median $= \dfrac{(57 + 79)}{2} = 68.0$, 20% trimmed mean = 66.2, 30% trimmed mean = 67.5.

Section 1.4

45.

    a. $\bar{x} = \frac{1}{n}\sum_i x_i = 577.9/5 = 115.58$. Deviations from the mean:

          116.4 - 115.58 = .82, 115.9 - 115.58 = .32, 114.6 -115.58 = -.98,
          115.2 - 115.58 = -.38, and 115.8-115.58 = .22.

    b. $s^2 = [(.82)^2 + (.32)^2 + (-.98)^2 + (-.38)^2 + (.22)^2]/(5-1) = 1.928/4 = .482$,
        so s = .694.

    c. $\sum_i x_i^2 = 66{,}795.61$, so $s^2 = \frac{1}{n-1}\left[\sum_i x_i^2 - \frac{1}{n}\left(\sum_i x_i\right)^2\right] =$

          $[66{,}795.61 - (577.9)^2/5]/4 = 1.928/4 = .482$.

47.   The sample mean, $\bar{x} = \frac{1}{n}\sum x_i = \frac{1}{10}(1{,}162) = \bar{x} = 116.2$.

The sample standard deviation,

$$s = \sqrt{\frac{\sum x_i^2 - \frac{\left(\sum x_i\right)^2}{n}}{n-1}} = \sqrt{\frac{140{,}992 - \frac{(1{,}162)^2}{10}}{9}} = 25.75$$

On average, we would expect a fracture strength of 116.2. In general, the size of a typical deviation from the sample mean (116.2) is about 25.75. Some observations may deviate from 116.2 by more than this and some by less..

49.

    a. $\Sigma x = 2.75 + ... + 3.01 = 56.80$,
        $\Sigma x^2 = (2.75)^2 + ... + (3.01)^2 = 197.8040$

    b. $s^2 = \dfrac{197.8040 - (56.80)^2/17}{16} = \dfrac{8.0252}{16} = .5016, \;\; s = .708$

51.

a. $\Sigma x = 2563$ and $\Sigma x^2 = 368,501$, so

$$s^2 = \frac{[368,501 - (2563)^2 / 19]}{18} = 1264.766 \text{ and } s = 35.564$$

b. If y = time in minutes, then y = cx where $c = \frac{1}{60}$, so

$$s_y^2 = c^2 s_x^2 = \frac{1264.766}{3600} = .351 \text{ and } s_y = cs_x = \frac{35.564}{60} = .593$$

53.

a. lower half: 2.34 2.43 2.62 2.74 2.74 2.75 2.78 3.01 3.46
upper half: 3.46 3.56 3.65 3.85 3.88 3.93 4.21 4.33 4.52
Thus the lower fourth is 2.74 and the upper fourth is 3.88.

b. $f_s = 3.88 - 2.74 = 1.14$

c. f_s wouldn't change, since increasing the two largest values does not affect the upper fourth.

d. By at most .40 (that is, to anything not exceeding 2.74), since then it will not change the lower fourth.

e. Since n is now even, the lower half consists of the smallest 9 observations and the upper half consists of the largest 9. With the lower fourth = 2.74 and the upper fourth = 3.93, $f_s = 1.19$.

55.

a. Lower half of the data set: 325   325   334   339   356   356   359   359   363   364   364   366   369, whose median, and therefore the lower quartile, is 359 (the 7th observation in the sorted list).
The top half of the data is 370   373   373   374   375   389   392   393   394   397   402   403   424, whose median, and therefore the upper quartile is 392. So, the IQR = 392 - 359 = 33.

b. 1.5(IQR) = 1.5(33) = 49.5 and 3(IQR) = 3(33) = 99. Observations that are further than 49.5 below the lower quartile (i.e., 359-49.5 = 309.5 or less) or more than 49.5 units above the upper quartile (greater than 392+49.5 = 441.5) are classified as 'mild' outliers. 'Extreme' outliers would fall 99 or more units below the lower, or above the upper, quartile. Since the minimum and maximum observations in the data are 325 and 424, we conclude that there are no mild outliers in this data (and therefore, no 'extreme' outliers either).

c. A boxplot (created by Minitab) of this data appears below. There is a slight positive skew to the data, but it is not far from being symmetric. The variation, however, seems large (the spread 424-325 = 99 is a large percentage of the median/typical value)

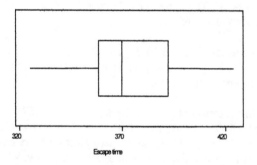

d. Not until the value x = 424 is lowered below the upper quartile value of 392.

e. Not until the value x = 424 is lowered below the upper quartile value of 392 would there be any change in the value of the upper quartile. That is, the value x = 424 could not be decreased by more than 424-392 = 32 units.

57.

a. 1.5(IQR) = 1.5(216.8-196.0) = 31.2 and 3(IQR) = 3(216.8-196.0) = 62.4. Mild outliers: observations below 196-31.2 = 164.6 or above 216.8+31.2 = 248. Extreme outliers: observations below 196-62.4 = 133.6 or above 216.8+62.4 = 279.2. Of the observations given, 125.8 is an extreme outlier and 250.2 is a mild outlier.

b. A boxplot of this data appears below. There is a bit of positive skew to the data but, except for the two outliers identified in part (a), the variation in the data is relatively small.

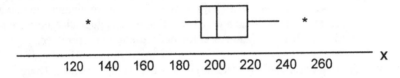

59.

a. ED: median = .4 (the 14th value in the *sorted* list of data). The lower quartile (median of the lower half of the data, including the median, since n is odd) is (.1+.1)/2 = .1. The upper quartile is (2.7+2.8)/2 = 2.75. Therefore, IQR = 2.75 - .1 = 2.65.

Non-ED: median = (1.5+1.7)/2 = 1.6. The lower quartile (median of the lower 25 observations) is .3; the upper quartile (median of the upper half of the data) is 7.9. Therefore, IQR = 7.9 - .3 = 7.6.

b. ED: mild outliers are less than .1 - 1.5(2.65) = -3.875 or greater than 2.75 + 1.5(2.65) = 6.725. Extreme outliers are less than .1 - 3(2.65) = -7.85 or greater than 2.75 + 3(2.65) = 10.7. So, the two largest observations (11.7, 21.0) are extreme outliers and the next two largest values (8.9, 9.2) are mild outliers. There are no outliers at the lower end of the data.

Non-ED: mild outliers are less than .3 - 1.5(7.6) = -11.1 or greater than 7.9 + 1.5(7.6) = 19.3. Note that there are no mild outliers in the data, hence there can not be any extreme outliers either.

c. A comparative boxplot appears below. The outliers in the ED data are clearly
visible. There is noticeable positive skewness in both samples; the Non-Ed data
has more variability then the Ed data; the typical values of the ED data tend to be
smaller than those for the Non-ED data.

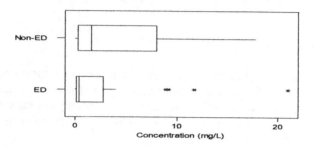

61. Outliers occur in the 6 a.m. data. The distributions at the other times are fairly
symmetric. Variability and the 'typical' values in the data increase a little at the 12
noon and 2 p.m. times.

Supplementary Exercises

63.

Flow rate	Median	Lower quartile	Upper quartile	IQR	1.5(IQR)	3(IQR)
125	3.1	2.7	3.8	1.1	1.65	.3
160	4.4	4.2	4.9	.7	1.05	.1
200	3.8	3.4	4.6	1.2	1.80	3.6

There are no outliers in the three data sets. However, as the comparative boxplot below shows, the three data sets differ with respect to their central values (the medians are different) and the data for flow rate 160 is somewhat less variable than the other data sets. Flow rates 125 and 200 also exhibit a small degree of positive skewness.

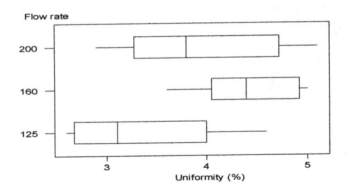

65.

a. HC data: $\sum_i x_i^2 = 2618.42$ and $\sum_i x_i = 96.8$,

so $s^2 = [2618.42 - (96.8)^2/4]/3 = 91.953$
and the sample standard deviation is $s = 9.59$.

CO data: $\sum_i x_i^2 = 145645$ and $\sum_i x_i = 735$, so $s^2 = [145645 - (735)^2/4]/3 = 3529.583$ and the sample standard deviation is $s = 59.41$.

b. The mean of the HC data is 96.8/4 = 24.2; the mean of the CO data is 735/4 = 183.75. Therefore, the coefficient of variation of the HC data is 9.59/24.2 = .3963, or 39.63%. The coefficient of variation of the CO data is 59.41/183.75 = .3233, or 32.33%. Thus, even though the CO data has a larger standard deviation than does the HC data, it actually exhibits *less* variability (in percentage terms) around its average than does the HC data.

67.

$$\sum x_i = 163.2$$

$$100\left(\frac{1}{15}\right)\%trimmedmean = \frac{163.2 - 8.5 - 15.6}{13} = 10.70$$

$$100\left(\frac{2}{15}\right)\%trimmedmean = \frac{163.2 - 8.5 - 8.8 - 15.6 - 13.7}{11} = 10.60$$

$$\therefore \frac{1}{2}(100)\left(\frac{1}{15}\right) + \frac{1}{2}(100)\left(\frac{2}{15}\right) = 100\left(\frac{1}{10}\right) = 10\%trimmedmean$$

$$= \frac{1}{2}(10.70) + \frac{1}{2}(10.60) = 10.65$$

69.

a.

$$\bar{y} = \frac{\sum y_i}{n} = \frac{\sum(ax_i + b)}{n} = \frac{a\sum x_i + b}{n} = a\bar{x} + b.$$

$$s_y^2 = \frac{\sum(y_i - \bar{y})^2}{n-1} = \frac{\sum(ax_i + b - (a\bar{x} + b))^2}{n-1} = \frac{\sum(ax_i - a\bar{x})^2}{n-1}$$

$$= \frac{a^2\sum(x_i - \bar{x})^2}{n-1} = a^2 s_x^2.$$

b.

$$x = {}^\circ C, y = {}^\circ F$$

$$\bar{y} = \frac{9}{5}(87.3) + 32 = 189.14$$

$$s_y = \sqrt{s_y^2} = \sqrt{\left(\frac{9}{5}\right)^2 (1.04)^2} = \sqrt{3.5044} = 1.872$$

71.

    a.   The mean, median, and trimmed mean are virtually identical, which suggests symmetry. If there are outliers, they are balanced. The range of values is only 25.5, but half of the values are between 132.95 and 138.25.

    b.

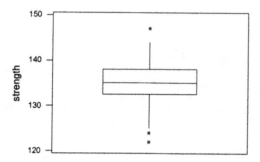

The boxplot also displays the symmetry, and adds a visual of the outliers, two on the lower end, and one on the upper.

73.

$$
\begin{array}{ll}
0.7\,|\,8 & \text{stem=tenths} \\
0.8\,|\,11556 & \text{leaf=hundredths} \\
0.9\,|\,2233335566 & \\
1.0\,|\,0566 & \\
\end{array}
$$

$\bar{x} = .9255, s = .0809, \tilde{x} = .93$

$lowerfourth = .855, upperfourth = .96$

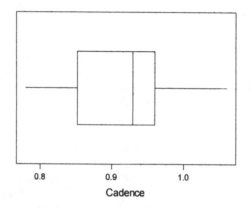

The data appears to be a bit skewed toward smaller values (negatively skewed). There are no outliers. The mean and the median are close in value.

75.

   a. The median is the same (371) in each plot and all three data sets are very symmetric. In addition, all three have the same minimum value (350) and same maximum value (392). Moreover, all three data sets have the same lower (364) and upper quartiles (378). So, all three boxplots will be *identical*.

b. A comparative dotplot is shown below. These graphs show that there are differences in the variability of the three data sets. They also show differences in the way the values are distributed in the three data sets.

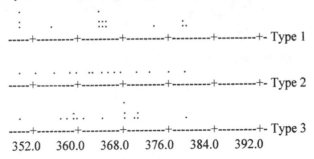

c. The boxplot in (a) is not capable of detecting the differences among the data sets. The primary reason is that boxplots give up some detail in describing data because they use only 5 summary numbers for comparing data sets. Note: The definition of lower and upper quartile used in this text is slightly different than the one used by some other authors (and software packages). Technically speaking, the median of the lower half of the data is not really the first quartile, although it is generally *very close*. Instead, the medians of the lower and upper halves of the data are often called the **lower** and **upper hinges**. Our boxplots use the lower and upper hinges to define the spread of the middle 50% of the data, but other authors sometimes use the *actual* quartiles for this purpose. The difference is usually very slight, usually unnoticeable, but not always. For example in the data sets of this exercise, a comparative boxplot based on the actual quartiles (as computed by Minitab) is shown below. The graph shows substantially the same type of information as those described in (a) except the graphs based on quartiles are able to detect the slight differences in variation between the three data sets.

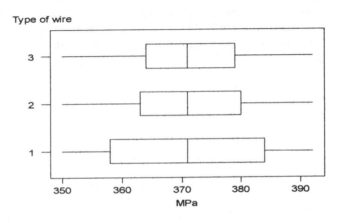

77.

a.

```
 0|2355566777888
 1|0000135555
 2|00257
 3|0033
 4|0057
 5|044
 6|                    stem: ones
 7|05                  leaf: tenths
 8|8
 9|0
10|3
HI|22.0 24.5
```

b.

Interval	Frequency	Rel. Freq.	Density
0 -< 2	23	.500	.250
2 -< 4	9	.196	.098
4 -< 6	7	.152	.076
6 -< 10	4	.087	.022
10 -< 20	1	.022	.002
20 -< 30	2	.043	.004

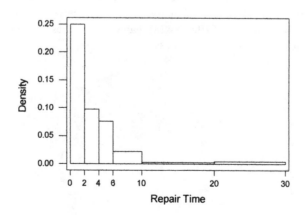

Chapter 1: Overview and Descriptive Statistics

79.

a. $\sum_{i=1}^{n+1} x_i = \sum_{i=1}^{n} x_i + x_{n+1} = n\bar{x}_n + x_{n+1}, so\, \bar{x}_{n+1} = \dfrac{[n\bar{x}_n + x_{n+1}]}{(n+1)}$

b.

$$ns_{n+1}^2 = \sum_{i=1}^{n+1}(x_i - \bar{x}_{n+1})^2 = \sum_{i=1}^{n+1} x_i^2 - (n+1)\bar{x}_{n+1}^2$$

$$= \sum_{i=1}^{n} x_i^2 - n\bar{x}_n^2 + x_{n+1}^2 + n\bar{x}_n^2 - (n+1)\bar{x}_{n+1}^2$$

$$= (n-1)s_n^2 + \left\{ x_{n+1}^2 + n\bar{x}_n^2 - (n+1)\bar{x}_{n+1}^2 \right\}$$

When the expression for $\bar{x}_{n+1}$ from **a** is substituted, the expression in braces simplifies to the following, as desired: $\dfrac{n(x_{n+1} - \bar{x}_n)^2}{(n+1)}$

81. Assuming that the histogram is unimodal, then there is evidence of positive skewness in the data since the median lies to the left of the mean (for a symmetric distribution, the mean and median would coincide). For more evidence of skewness, compare the distances of the 5th and 95th percentiles from the median: median - 5th percentile = 500 - 400 = 100 while 95th percentile -median = 720 - 500 = 220. Thus, the largest 5% of the values (above the 95th percentile) are further from the median than are the lowest 5%. The same skewness is evident when comparing the 10th and 90th percentiles to the median: median - 10th percentile = 500 - 430 = 70 while 90th percentile -median = 640 - 500 = 140. Finally, note that the largest value (925) is much further from the median (925-500 = 425) than is the smallest value (500 - 220 = 280), again an indication of positive skewness.

83.

a. When there is perfect symmetry, the smallest observation y_1 and the largest observation y_n will be equidistant from the median, so $y_n - \bar{x} = \bar{x} - y_1$.

   Similarly, the second smallest and second largest will be equidistant from the median, so $y_{n-1} - \bar{x} = \bar{x} - y_2$

   and so on. Thus, the first and second numbers in each pair will be equal, so that each point in the plot will fall exactly on the 45 degree line. When the data is positively skewed, y_n will be much further from the median than is y_1, so $y_n - \tilde{x}$ will considerably exceed $\tilde{x} - y_1$ and the point $(y_n - \tilde{x}, \tilde{x} - y_1)$ will fall considerably below the 45 degree line. A similar comment aplies to other points in the plot.

b. The first point in the plot is $(2745.6 - 221.6, 221.6\ 0- 4.1) = (2524.0, 217.5)$. The others are: (1476.2, 213.9), (1434.4, 204.1), (756.4, 190.2), (481.8, 188.9), (267.5, 181.0), (208.4, 129.2), (112.5, 106.3), (81.2, 103.3), (53.1, 102.6), (53.1, 92.0), (33.4, 23.0), and (20.9, 20.9). The first number in each of the first seven pairs greatly exceed the second number, so each point falls well below the 45 degree line. A substantial positive skew (stretched upper tail) is indicated.

Chapter 1: Overview and Descriptive Statistics

CHAPTER 2

Section 2.1

1.

   a. S = { 1324, 1342, 1423, 1432, 2314, 2341, 2413, 2431, 3124, 3142, 4123, 4132, 3214, 3241, 4213, 4231 }

   b. Event A contains the outcomes where 1 is first in the list:
      A = { 1324, 1342, 1423, 1432 }

   c. Event B contains the outcomes where 2 is first or second:
      B = { 2314, 2341, 2413, 2431, 3214, 3241, 4213, 4231 }

   d. The compound event A∪B contains the outcomes in A or B or both:
      A∪B = {1324, 1342, 1423, 1432, 2314, 2341, 2413, 2431, 3214, 3241, 4213, 4231 }

3.

   a. Event A = { SSF, SFS, FSS }

   b. Event B = { SSS, SSF, SFS, FSS }

   c. For Event C, the system must have component 1 working (S in the first position), then at least one of the other two components must work (at least one S in the 2nd and 3rd positions: Event C = { SSS, SSF, SFS }

   d. Event C′ = { SFF, FSS, FSF, FFS, FFF }
      Event A∪C = { SSS, SSF, SFS, FSS }
      Event A∩C = { SSF, SFS }
      Event B∪C = { SSS, SSF, SFS, FSS }
      Event B∩C = { SSS SSF, SFS }

5.

a.

Outcome Number	Outcome	Outcome Number	Outcome
1	111	15	223
2	112	16	231
3	113	17	232
4	121	18	233
5	122	19	311
6	123	20	312
7	131	21	313
8	132	22	321
9	133	23	322
10	211	24	323
11	212	25	331
12	213	26	332
13	221	27	333
14	222		

b. Outcome Numbers 1, 14, 27

c. Outcome Numbers 6, 8, 12, 16, 20, 22

d. Outcome Numbers 1, 3, 7, 9, 19, 21, 25, 27

7.

a. S = {BBBAAAA, BBABAAA, BBAABAA, BBAAABA, BBAAAAB, BABBAAA, BABABAA, BABAABA, BABAAAB, BAABBAA, BAABABA, BAABAAB, BAAABBA, BAAABAB, BAAAABB, ABBBAAA, ABBABAA, ABBAABA, ABBAAAB, ABABBAA, ABABABA, ABABAAB, ABAABBA, ABAABAB, ABAAABB, AABBBAA, AABBABA, AABBAAB, AABABBA, AABABAB, AABAABB, AAABBBA, AAABBAB, AAABABB, AAAABBB}

b. {AAAABBB, AAABABB, AAABBAB, AABAABB, AABABAB}

9.

    a.   In the diagram on the left, the shaded area is $(A \cup B)'$. On the right, the shaded area is A', the striped area is B', and the intersection $A' \cap B'$ occurs where there is BOTH shading and stripes. These two diagrams display the same area.

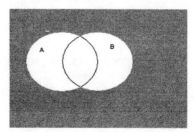 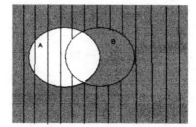

    b.   In the diagram below, the shaded area represents $(A \cap B)'$. Using the diagram on the right above, the union of A' and B' is represented by the areas that have either shading or stripes or both. Both of the diagrams display the same area.

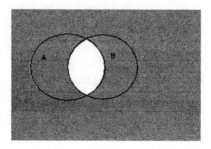 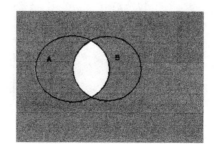

Section 2.2

11.

    a.   .07

    b. $.15 + .10 + .05 = .30$

    c.   Let event A = selected customer owns stocks. Then the probability that a selected customer does not own a stock can be represented by $P(A') = 1 - P(A) = 1 - (.18 + .25) = 1 - .43 = .57$. This could also have been done easily by adding the probabilities of the funds that are not stocks.

13.

    a. awarded either #1 or #2 (or both):

$$P(A_1 \cup A_2) = P(A_1) + P(A_2) - P(A_1 \cap A_2) = .22 + .25 - .11 = .36$$

    b. awarded neither #1 or #2:

$$P(A_1' \cap A_2') = P[(A_1 \cup A_2)\,'] = 1 - P(A_1 \cup A_2) = 1 - .36 = .64$$

    c. awarded at least one of #1, #2, #3:

$$P(A_1 \cup A_2 \cup A_3) = P(A_1) + P(A_2) + P(A_3) - P(A_1 \cap A_2) - P(A_1 \cap A_3) -$$
$$P(A_2 \cap A_3) + P(A_1 \cap A_2 \cap A_3)$$
$$= .22 + .25 + .28 - .11 - .05 - .07 + .01 = .53$$

    d. awarded none of the three projects:

$$P(A_1' \cap A_2' \cap A_3') = 1 - P(\text{awarded at least one}) = 1 - .53 = .47.$$

    e. awarded #3 but neither #1 nor #2:

$$P(A_1' \cap A_2' \cap A_3) = P(A_3) - P(A_1 \cap A_3) - P(A_2 \cap A_3)$$
$$+ P(A_1 \cap A_2 \cap A_3)$$
$$= .28 - .05 - .07 + .01 \quad = .17$$

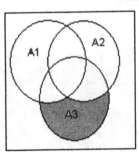

f. either (neither #1 nor #2) or #3: $P[(A_1' \cap A_2') \cup A_3]$
= P(shaded region) = P(awarded none) + $P(A_3)$ = .47 + .28 = .75. Alternatively,
answers to **a** – **f** can be obtained from probabilities on the Venn diagram on the
right

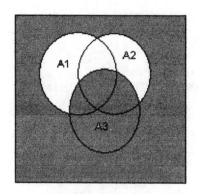 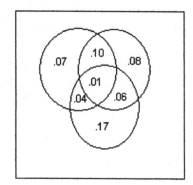

15.

a. Let event E be the event that at most one purchases an electric dryer. Then E' is
the event that at least two purchase electric dryers.
$P(E') = 1 - P(E) = 1 - .428 = .572$

b. Let event A be the event that all five purchase gas. Let event B be the event that
all five purchase electric. All other possible outcomes are those in which at least
one of each type is purchased. Thus, the desired probability =
$1 - P(A) - P(B) = 1 - .116 - .005 = .879$

17.

a. The probabilities do not add to 1 because there are other software packages
besides SPSS and SAS for which requests could be made.

b. $P(A') = 1 - P(A) = 1 - .30 = .70$

c. $P(A \cup B) = P(A) + P(B) = .30 + .50 = .80$
    (since A and B are mutually exclusive events)

d. $P(A' \cap B') = P[(A \cup B)']$ (De Morgan's law)
    $= 1 - P(A \cup B)$
    $= 1 - .80 = .20$

19.   Let event A be that the selected joint was found defective by inspector A. $P(A) = \frac{724}{10,000}$. Let event B be analogous for inspector B. $P(B) = \frac{751}{10,000}$. Compound event $A \cup B$ is the event that the selected joint was found defective by at least one of the two inspectors. $P(A \cup B) = \frac{1159}{10,000}$.

    a.   The desired event is $(A \cup B)'$, so we use the complement rule:
$$P(A \cup B)' = 1 - P(A \cup B) = 1 - \frac{1159}{10,000} = \frac{8841}{10,000} = .8841$$

    b.   The desired event is $B \cap A'$. $P(B \cap A') = P(B) - P(A \cap B)$.
$P(A \cap B) = P(A) + P(B) - P(A \cup B), \; = .0724 + .0751 - .1159 = .0316$
So $P(B \cap A') = P(B) - P(A \cap B) = .0751 - .0316 = .0435$

21.

    a. $P(\{M,H\}) = .10$

    b. $P(\text{low auto}) = P[\{(L,N), (L,L), (L,M), (L,H)\}] = .04 + .06 + .05 + .03 = .18$
Following a similar pattern, $P(\text{low homeowner's}) = .06 + .10 + .03 = .19$

    c. $P(\text{same deductible for both}) = P[\{ LL, MM, HH \}] = .06 + .20 + .15 = .41$

    d. $P(\text{deductibles are different}) = 1 - P(\text{same deductibles}) = 1 - .41 = .59$

    e. $P(\text{at least one low deductible}) = P[\{ LN, LL, LM, LH, ML, HL \}]$
$$= .04 + .06 + .05 + .03 + .10 + .03 = .31$$

    f. $P(\text{neither low}) = 1 - P(\text{at least one low}) = 1 - .31 = .69$

23. Assume that the computers are numbered 1 – 6 as described. Also assume that computers 1 and 2 are the laptops. Possible outcomes are (1,2) (1,3) (1,4) (1,5) (1,6) (2,3) (2,4) (2,5) (2,6) (3,4) (3,5) (3,6) (4,5) (4,6) and (5,6).

    a. P(both are laptops) = P[{ (1,2)}] = $\frac{1}{15}$ = .067

    b. P(both are desktops) = P[{(3,4) (3,5) (3,6) (4,5) (4,6) (5,6)}] = $\frac{6}{15}$ = .40

    c. P(at least one desktop) = 1 – P(no desktops)
        = 1 – P(both are laptops)   = 1 – .067 = .933

    d. P(at least one of each type) = 1 – P(both are the same)
        = 1 – P(both laptops) – P(both desktops) = 1 - .067 - .40 = .533

25. $P(A \cap B) = P(A) + P(B) - P(A \cup B) = .65$
$P(A \cap C) = .55, \ P(B \cap C) = .60$
$P(A \cap B \cap C) = P(A \cup B \cup C) - P(A) - P(B) - P(C)$
          $+ P(A \cap B) + P(A \cap C) + P(B \cap C)$
        $= .98 - .7 - .8 - .75 + .65 + .55 + .60$
        $= .53$

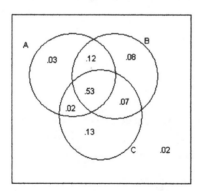

    a. $P(A \cup B \cup C) = .98$, as given.

    b. P(none selected) = 1 - $P(A \cup B \cup C)$ = 1 - .98 = .02

    c. P(only automatic transmission selected) = .03 from the Venn Diagram

    d. P(exactly one of the three) = .03 + .08 + .13 = .24

27. Outcomes: (A,B) (A,C_1) (A,C_2) (A,F) (B,A) (B,C_1) (B,C_2) (B,F)
(C₁,A) (C₁,B) (C₁,C₂) (C₁,F) (C₂,A) (C₂,B) (C₂,C₁) (C₂,F)
(F,A) (F,B) (F,C₁) (F,C₂)

a. $P[(A,B) \text{ or } (B,A)] = \frac{2}{20} = \frac{1}{10} = .1$

b. $P(\text{at least one C}) = \frac{14}{20} = \frac{7}{10} = .7$

c. $P(\text{at least 15 years}) = 1 - P(\text{at most 14 years})$
$= 1 - P[(3,6) \text{ or } (6,3) \text{ or } (3,7) \text{ or } (7,3) \text{ or } (3,10) \text{ or } (10,3) \text{ or } (6,7) \text{ or } (7,6)]$
$= 1 - \frac{8}{20} = 1 - .4 = .6$

Section 2.3

29.

a. $(5)(4) = 20$ (5 choices for president, 4 remain for vice president)

b. $(5)(4)(3) = 60$

c. $\binom{5}{2} = \frac{5!}{2!3!} = 10$ (No ordering is implied in the choice)

31.

a. $(n_1)(n_2) = (9)(27) = 243$

b. $(n_1)(n_2)(n_3) = (9)(27)(15) = 3645$, so such a policy could be carried out for 3645 successive nights, or approximately 10 years, without repeating exactly the same program.

33.

a. $\dbinom{20}{5} = \dfrac{20!}{5! \, 15!} = 15{,}504$

b. $\dbinom{8}{4} \cdot \dbinom{12}{1} = 840$

c. P(exactly 4 have cracks) $= \dfrac{\dbinom{8}{4}\dbinom{12}{1}}{\dbinom{20}{5}} = \dfrac{840}{15{,}504} = .0542$

d. P(at least 4) = P(exactly 4) + P(exactly 5)

$$= \frac{\dbinom{8}{4}\dbinom{12}{1}}{\dbinom{20}{5}} + \frac{\dbinom{8}{5}\dbinom{12}{0}}{\dbinom{20}{5}} = .0542 + .0036 = .0578$$

35.   There are 10 possible outcomes -- $\dbinom{5}{2}$ ways to select the positions for B's votes:

BBAAA, BABAA, BAABA, BAAAB, ABBAA, ABABA, ABAAB, AABBA, AABAB, and AAABB. Only the last two have A ahead of B throughout the vote count. Since the outcomes are equally likely, the desired probability is $\frac{2}{10} = .20$.

37.   There are $\dbinom{60}{5}$ ways to select the 5 runs. Each catalyst is used in 12 different runs, so the number of ways of selecting one run from each of these 5 groups is 12^5. Thus the desired probability is $\dfrac{12^5}{\dbinom{60}{5}} = .0456$.

39.

a. We want to choose all of the 5 cordless, and 5 of the 10 others, to be among the

first 10 serviced, so the desired probability is $\dfrac{\dbinom{5}{5}\dbinom{10}{5}}{\dbinom{15}{10}} = \dfrac{252}{3003} = .0839$

b. Isolating one group, say the cordless phones, we want the other two groups represented in the last 5 serviced. So we choose 5 of the 10 others, except that we don't want to include the outcomes where the last five are all the same.

So we have $\dfrac{\dbinom{10}{5} - 2}{\dbinom{15}{5}}$. But we have three groups of phones, so the desired

probability is $\dfrac{3 \cdot \left[\dbinom{10}{5} - 2\right]}{\dbinom{15}{5}} = \dfrac{3(250)}{3003} = .2498$.

c. We want to choose 2 of the 5 cordless, 2 of the 5 cellular, and 2 of the corded

phones: $\dfrac{\dbinom{5}{2}\dbinom{5}{2}\dbinom{5}{2}}{\dbinom{15}{6}} = \dfrac{1000}{5005} = .1998$

41.

a. P(at least one F among 1st 3) = 1 – P(no F's among 1st 3)

$$= 1 - \frac{4 \times 3 \times 2}{8 \times 7 \times 6} = 1 - \frac{24}{336} = 1 - .0714 = .9286$$

An alternative method to calculate P(no F's among 1st 3)
would be to choose none of the females and 3 of the 4 males, as follows:

$$\frac{\binom{4}{0}\binom{4}{3}}{\binom{8}{3}} = \frac{4}{56} = .0714, \text{ obviously producing the same result.}$$

b. P(all F's among 1st 5) = $\dfrac{\binom{4}{4}\binom{4}{1}}{\binom{8}{5}} = \dfrac{4}{56} = .0714$

c. P(orderings are different) = 1 – P(orderings are the same for both semesters)
= 1 – (# orderings such that the orders are the same each semester)/(total # of
possible orderings for 2 semesters)

$$= 1 - \frac{8 \times 7 \times 6 \times 5 \times 4 \times 3 \times 2 \times 1}{(8 \times 7 \times 6 \times 5 \times 4 \times 3 \times 2 \times 1 \times) \times (8 \times 7 \times 6 \times 5 \times 4 \times 3 \times 2 \times 1)}$$

$$= .99997520$$

43.   # of 10 high straights = 4×4×4×4×4 (4 – 10's, 4 – 9's , etc)

$$P(\text{10 high straight}) = \frac{4^5}{\binom{52}{5}} = \frac{1024}{2,598,960} = .000394$$

$$P(\text{straight}) = 10 \times \frac{4^5}{\binom{52}{5}} = .003940 \text{ (Multiply by 10 because there are 10 different}$$

card values that could be high: Ace, King, etc.) There are only 40 straight flushes (10 in each suit), so

$$P(\text{straight flush}) = \frac{40}{\binom{52}{5}} = .00001539$$

Section 2.4

45.

    a.   P(A) = .106 + .141 + .200 = .447, P(C) =.215 + .200 + .065 + .020 = .500 P(A ∩ C) = .200

    b. $P(A|C) = \dfrac{P(A \cap C)}{P(C)} = \dfrac{.200}{.500} = .400$. If we know that the individual came from ethnic group 3, the probability that he has type A blood is .40. $P(C|A) = \dfrac{P(A \cap C)}{P(A)} = \dfrac{.200}{.447} = .447$. If a person has type A blood, the probability that he is from ethnic group 3 is .447

    c.   Define event D = {ethnic group 1 selected}. We are asked for P(D|B′) = $\dfrac{P(D \cap B')}{P(B')} = \dfrac{.200}{.500} = .400$. P(D∩B′)=.082 + .106 + .004 = .192, P(B′) = 1 – P(B) = 1 – [.008 + .018 + .065] = .909

47.

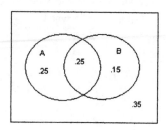

a. P(B│A) =
$$\frac{P(A \cap B)}{P(A)} = \frac{.25}{.50} = .50$$

c. P(A│B) =
$$\frac{P(A \cap B)}{P(B)} = \frac{.25}{.40} = .6125$$

b. P(B′│A) =
$$\frac{P(A \cap B')}{P(A)} = \frac{.25}{.50} = .50$$

d. P(A′│B) =
$$\frac{P(A' \cap B)}{P(B)} = \frac{.15}{.40} = .3875$$

e. P(A│A∪B) = $\dfrac{P[A \cap (A \cup B)]}{P(A \cup B)} = \dfrac{.50}{.65} = .7692$

49. The first desired probability is P(both bulbs are 75 watt | at least one is 75 watt).

P(at least one is 75 watt) = 1 – P(none are 75 watt)

$$= 1 - \frac{\binom{9}{2}}{\binom{15}{2}} = 1 - \frac{36}{105} = \frac{69}{105}.$$

Notice that P[(both are 75 watt)∩(at least one is 75 watt)]

$$= \text{P(both are 75 watt)} = \frac{\binom{6}{2}}{\binom{15}{2}} = \frac{15}{105}.$$

So P(both bulbs are 75 watt | at least one is 75 watt) = $\dfrac{\dfrac{15}{105}}{\dfrac{69}{105}} = \dfrac{15}{69} = .2174$

Second, we want P(same rating | at least one NOT 75 watt).

P(at least one NOT 75 watt) = 1 – P(both are 75 watt)

$$= 1 - \frac{15}{105} = \frac{90}{105}.$$

Now, P[(same rating)∩(at least one not 75 watt)] = P(both 40 watt or both 60 watt).

$$\text{P(both 40 watt or both 60 watt)} = \frac{\binom{4}{2} + \binom{5}{2}}{\binom{15}{2}} = \frac{16}{105}$$

Now, the desired conditional probability is $\dfrac{\dfrac{16}{105}}{\dfrac{90}{105}} = \dfrac{16}{90} = .1778$

51.

a. P(R from 1st ∩ R from 2nd) = P(R from 2nd | R from 1st) • P(R from 1st)

$$= \frac{8}{11} \bullet \frac{6}{10} = .436$$

b. P(same numbers)   = P(both selected balls are the same color)

$$= P(\text{both red}) + P(\text{both green}) = .436 + \frac{4}{11} \bullet \frac{4}{10} = .581$$

53. $P(B \mid A) = \dfrac{P(A \cap B)}{P(A)} = \dfrac{P(B)}{P(A)}$ (since B is contained in A, A ∩ B = B)

$$= \frac{.05}{.60} = .0833$$

55.

a. P(A B) = P(B|A)•P(A) = $\dfrac{2 \times 1}{4 \times 3} \times \dfrac{2 \times 1}{6 \times 5} = .0111$

b. P(two other H's next to their wives | J and M together in the middle)

$$\frac{P[(H - W.or.W - H)and(J - M.or.M - J)and(H - W.or.W - H)]}{P(J - M.or.M - J.in.the.middle)}$$

numerator = $\dfrac{4 \times 1 \times 2 \times 1 \times 2 \times 1}{6 \times 5 \times 4 \times 3 \times 2 \times 1} = \dfrac{16}{6!}$

denominator = $\dfrac{4 \times 3 \times 2 \times 1 \times 2 \times 1}{6 \times 5 \times 4 \times 3 \times 2 \times 1} = \dfrac{48}{6!}$

so the desired probability = $\dfrac{16}{48} = \dfrac{1}{3}$.

c. P(all H's next to W's | J & M together)
  = P(all H's next to W's – including J&M)/P(J&M together)

$$= \frac{\dfrac{6 \times 1 \times 4 \times 1 \times 2 \times 1}{6!}}{\dfrac{5 \times 2 \times 1 \times 4 \times 3 \times 2 \times 1}{6!}} = \frac{48}{240} = .2$$

57.
$$P(A|B) + P(A'|B) = \frac{P(A \cap B)}{P(B)} + \frac{P(A' \cap B)}{P(B)}$$
$$= \frac{P(A \cap B) + P(A' \cap B)}{P(B)} = \frac{P(B)}{P(B)} = 1$$

59.

$.4 \times .3 = .12 = P(A_1 \cap B) = P(A_1) \bullet P(B|A)$

$.35 \times .6 = .21 = P(A_2 \cap B)$

$.25 \times .5 = .125 = P(A_3 \cap B)$

a. $P(A_2 \cap B) = .21$

b. $P(B) = P(A_1 \cap B) + P(A_2 \cap B) + P(A_3 \cap B) = .455$

c. $P(A_1|B) = \dfrac{P(A_1 \cap B)}{P(B)} = \dfrac{.12}{.455} = .264$

$P(A_2|B) = \dfrac{.21}{.455} = .462, \ P(A_3|B) = 1 - .264 - .462 = .274$

61. P(0 def in sample | 0 def in batch) = 1

$$P(0 \text{ def in sample} \mid 1 \text{ def in batch}) = \frac{\binom{9}{2}}{\binom{10}{2}} = .800$$

$$P(1 \text{ def in sample} \mid 1 \text{ def in batch}) = \frac{\binom{9}{1}}{\binom{10}{2}} = .200$$

$$P(0 \text{ def in sample} \mid 2 \text{ def in batch}) = \frac{\binom{8}{2}}{\binom{10}{2}} = .622$$

$$P(1 \text{ def in sample} \mid 2 \text{ def in batch}) = \frac{\binom{2}{1}\binom{8}{1}}{\binom{10}{2}} = .356$$

$$P(2 \text{ def in sample} \mid 2 \text{ def in batch}) = \frac{1}{\binom{10}{2}} = .022$$

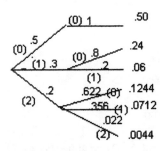

a. P(0 def in batch | 0 def in sample) = $\dfrac{.5}{.5+.24+.1244}$ = .578

P(1 def in batch | 0 def in sample) = $\dfrac{.24}{.5+.24+.1244}$ = .278

P(2 def in batch | 0 def in sample) = $\dfrac{.1244}{.5+.24+.1244}$ = .144

b. P(0 def in batch | 1 def in sample) = 0

P(1 def in batch | 1 def in sample) = $\dfrac{.06}{.06+.0712}$ = .457

P(2 def in batch | 1 def in sample) = $\dfrac{.0712}{.06+.0712}$ = .543

63.

a.

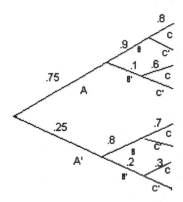

b. P(A $\cap$ B $\cap$ C) = .75 $\times$ .9 $\times$ .8 = .5400

c. P(B $\cap$ C) = P(A $\cap$ B $\cap$ C) + P(A' $\cap$ B $\cap$ C) = .5400+.25$\times$.8$\times$.7 = .6800

d. P(C) = P(A $\cap$ B $\cap$ C)+P(A' $\cap$ B $\cap$ C) + P(A $\cap$ B' $\cap$ C) + P(A' $\cap$ B' $\cap$ C)
    = .54+.045+.14+.015 = .74

e. P(A|B $\cap$ C) = $\dfrac{P(A\cap B\cap C)}{P(B\cap C)}$ = $\dfrac{.54}{.68}$ = .7941

65.

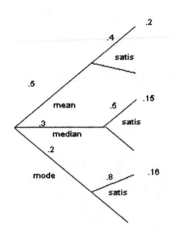

P(satis) = .51

$$P(\text{mean} \mid \text{satis}) = \frac{.2}{.51} = .3922$$

P(median | satis) = .2941

P(mode | satis) = .3137

So Mean (and not Mode!) is the most
likely author, while Median is least.

67.

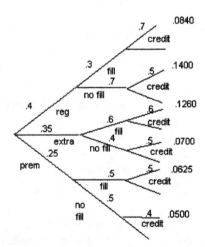

a. P(U ∩ F ∩ Cr) = .1260

b. P(Pr ∩ NF ∩ Cr) = .05

c. P(Pr ∩ Cr) = .0625 + .05
= .1125

d. P(F ∩ Cr)= .0840+ .1260 + .0625 = .2725

e. P(Cr) = .5325

f. P(PR | Cr) =
$$\frac{P(\text{Pr} \cap Cr)}{P(Cr)} = \frac{.1125}{.5325} = .2113$$

Section 2.5

69.

    a.   Since the events are independent, then A′ and B′ are independent, too. (see paragraph below equation 2.7. $P(B'|A') = . P(B') = 1 - .7 = .3$

    b. $P(A \cup B) = P(A) + P(B) - P(A) \cdot P(B) = .4 + .7 + (.4)(.7) = .82$

    c. $P(AB' | A \cup B) = \dfrac{P(AB' \cap (A \cup B))}{P(A \cup B)} = \dfrac{P(AB')}{P(A \cup B)} = \dfrac{.12}{.82} = .146$

71. $P(A' \cap B) = P(B) - P(A \cap B) = P(B) - P(A) \bullet P(B) = [1 - P(A)] \bullet P(B) = P(A') \bullet P(B)$.

    Alternatively, $P(A' | B) = \dfrac{P(A' \cap B)}{P(B)} = \dfrac{P(B) - P(A \cap B)}{P(B)}$

$$= \dfrac{P(B) - P(A) \cdot P(B)}{P(B)} = 1 - P(A) = P(A').$$

73.   Let event E be the event that an error was signaled incorrectly. We want P(at least one signaled incorrectly) $= P(E_1 \cup E_2 \cup ... \cup E_{10}) = 1 - P(E_1' \cap E_2' \cap ... \cap E_{10}')$. $P(E') = 1 - .05 = .95$. For 10 independent points, $P(E_1' \cap E_2' \cap ... \cap E_{10}') = P(E_1')P(E_2')...P(E_{10}')$ so $= P(E_1 \cup E_2 \cup ... \cup E_{10}) = 1 - [.95]^{10} = .401$. Similarly, for 25 points, the desired probability is $= 1 - [P(E')]^{25} = 1 - (.95)^{25} = .723$

75.   Let q denote the probability that a rivet is defective.

    a.   P(seam need rework) $= .20 = 1 - $ P(seam doesn't need rework)

                  $= 1 - $ P(no rivets are defective)

                  $= 1 - $ P(1st isn't def $\cap ... \cap 25^{th}$ isn't def)

                  $= 1 - (1 - q)^{25}$, so $.80 = (1 - q)^{25}$, $1 - q = (.80)^{1/25}$, and thus q $= 1 - .99111 = .00889$.

    b.   The desired condition is $.10 = 1 - (1 - q)^{25}$, i.e. $(1 - q)^{25} = .90$, from which q $= 1 - .99579 = .00421$.

77.   Let A_1 = older pump fails, A_2 = newer pump fails, and $x = P(A_1 \cap A_2)$. Then $P(A_1) = .10$
+ x, $P(A_2) = .05 + x$, and $x = P(A_1 \cap A_2) = P(A_1) \bullet P(A_2) = (.10 + x)(.05 + x)$. The
resulting quadratic equation, $x^2 - .85x + .005 = 0$, has roots x = .0059 and x = .8441.
Hopefully the smaller root is the actual probability of system failure.

79.

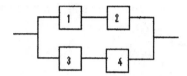

Using the hints, let $P(A_i) = p$, and $x = p^2$, then P(system lifetime exceeds t_0) $= p^2 + p^2 -$
$p^4 = 2p^2 - p^4 = 2x - x^2$. Now, set this equal to .99, or $2x - x^2 = .99 \Rightarrow x^2 - 2x + .99 = 0$.

Use the quadratic formula to solve for x: $= \dfrac{2 \pm \sqrt{4 - (4)(.99)}}{2} = \dfrac{2 \pm .2}{2} = 1 \pm .1 =$

.99 or 1.01 Since the value we want is a probability, and has to be ≤ 1, we use the value
of .99.

81.   P(both detect the defect) = 1 – P(at least one doesn't) = 1 - .2 = .8

a.   P(1st detects $\cap$ 2nd doesn't) = P(1st detects) – P(1st does $\cap$ 2nd does)
= .9 - .8 = .1
Similarly, P(1st doesn't $\cap$ 2nd does) = .1, so P(exactly one does)= .1+.1= .2

b.   P(neither detects a defect) = 1 – [P(both do) + P(exactly 1 does)]
= 1 – [.8+.2] = 0
so P(all 3 escape) = (0)(0)(0) = 0.

83.

a.   Let D_1 = detection on 1st fixation, D_2 = detection on 2nd fixation.
P(detection in at most 2 fixations) = $P(D_1) + P(D_1' \cap D_2)$
$= P(D_1) + P(D2 \mid D1')P(D_1)$
$= p + p(1 - p) = p(2 - p)$.

b. Define $D_1, D_2, \ldots, D_n$ as in **a**. Then P(at most n fixations)

$= P(D_1) + P(D_1' \cap D_2) + P(D_1' \cap D_2' \cap D_3) + \ldots + P(D_1' \cap D_2' \cap \ldots \cap D_{n-1}' \cap D_n)$

$= p + p(1-p) + p(1-p)^2 + \ldots + p(1-p)^{n-1}$

$= p[1 + (1-p) + (1-p)^2 + \ldots + (1-p)^{n-1}] = p \bullet \dfrac{1-(1-p)^n}{1-(1-p)} = 1-(1-p)^n$

Alternatively, P(at most n fixations) $= 1 - $ P(at least n+1 are req'd)

$\qquad\qquad\qquad\qquad\qquad\qquad = 1 - $ P(no detection in 1st n fixations)

$\qquad\qquad\qquad\qquad\qquad\qquad = 1 - P(D_1' \cap D_2' \cap \ldots \cap D_n')$

$\qquad\qquad\qquad\qquad\qquad\qquad = 1 - (1-p)^n$

c. P(no detection in 3 fixations) $= (1-p)^3$

d. P(passes inspection) $= $ P({not flawed} $\cup$ {flawed and passes})

$\qquad\qquad\qquad\qquad\qquad = $ P(not flawed) + P(flawed and passes)

$\qquad\qquad\qquad\qquad\qquad = .9 + $ P(passes | flawed)$\bullet$ P(flawed) $= .9+(1-p)^3(.1)$

e. P(flawed | passed) $= \dfrac{P(\textit{flawed} \cap \textit{passed})}{P(\textit{passed})} = \dfrac{.1(1-p)^3}{.9+.1(1-p)^3}$

$\qquad$ For p = .5, P(flawed | passed) $= \dfrac{.1(.5)^3}{.9+.1(.5)^3} = .0137$

85. P(system works) $= $ P(1 – 2 works $\cap$ 3 – 4 – 5 – 6 works $\cap$ 7 works)

$\qquad\qquad\qquad\qquad = $ P(1 – 2 works) $\bullet$ P(3 – 4 – 5 – 6 works) $\bullet$P(7 works)

$\qquad\qquad\qquad\qquad = (.99)(.9639)(.9) = .8588$

With the subsystem in figure 2.14 connected in parallel to this subsystem,

P(system works) $= .8588+.927 - (.8588)(.927) = .9897$

87.

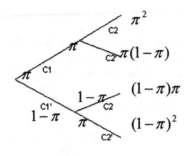

$1 - \pi$

P(at most 1 is lost) = 1 − P(both lost)

$$= 1 - \pi^2$$

P(exactly 1 lost) = $2\pi(1 - \pi)$

P(exactly 1 | at most 1) = $\dfrac{P(exactly 1)}{P(at.most 1)} = \dfrac{2\pi(1 - \pi)}{1 - \pi^2}$

Supplementary Exercises

89.

a. P(line 1) = $\dfrac{500}{1500} = .333$;

P(Crack) = $\dfrac{.50(500) + .44(400) + .40(600)}{1500} = \dfrac{666}{1500} = .444$

b. P(Blemish | line 1) = .15

c. P(Surface Defect) = $\dfrac{.10(500) + .08(400) + .15(600)}{1500} = \dfrac{172}{1500}$

P(line 1 and Surface Defect) = $\dfrac{.10(500)}{1500} = \dfrac{50}{1500}$

So P(line 1 | Surface Defect) = $= \dfrac{50/1500}{172/1500} = .291$

91. $P(A \cup B) = P(A) + P(B) - P(A)P(B)$
             $.626 \ = P(A) + P(B) - .144$

So $P(A) + P(B) = .770$ and $P(A)P(B) = .144$.
Let $x = P(A)$ and $y = P(B)$, then using the first equation, $y = .77 - x$, and substituting this into the second equation, we get $x(.77 - x) = .144$ or
$x^2 - .77x + .144 = 0$. Use the quadratic formula to solve:

$$\frac{.77 \pm \sqrt{.77^2 - (4)(.144)}}{2} = \frac{.77 \pm .13}{2} = .32 \text{ or } .45$$

So $P(A) = .45$ and $P(B) = .32$

93.

   a.   There are $5 \times 4 \times 3 \times 2 \times 1 = 120$ possible orderings, so $P(BCDEF) = \frac{1}{120} = .0083$

   b.   # orderings in which F is $3^{rd} = 4 \times 3 \times 1 * \times 2 \times 1 = 24$, (* because F must be here), so
       $P(F \ 3^{rd}) = \frac{24}{120} = .2$

   c. $P(F \text{ last}) = \dfrac{4 \times 3 \times 2 \times 1 \times 1}{120} = .2$

95.   When three experiments are performed, there are 3 different ways in which detection can occur on exactly 2 of the experiments: (i) #1 and #2 and not #3 (ii) #1 and not #2 and #3; (iii) not#1 and #2 and #3. If the impurity is present, the probability of exactly 2 detections in three (independent) experiments is $(.8)(.8)(.2) + (.8)(.2)(.8) + (.2)(.8)(.8) = .384$. If the impurity is absent, the analogous probability is $3(.1)(.1)(.9) = .027$. Thus
P(present | detected in exactly 2 out of 3) =

$$\frac{P(\det ected.in.exactly.2 \cap present)}{P(\det ected.in.exactly.2)}$$

$$= \frac{(.384)(.4)}{(.384)(.4) + (.027)(.6)} = .905$$

97.

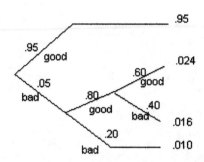

a. P(pass inspection) = P(pass initially ∪ passes after recrimping) = P(pass initially) + P(fails initially ∩ goes to recrimping ∩ is corrected after recrimping)
= .95 + (.05)(.80)(.60) (following path "bad-good-good" on tree diagram)
= .974

b. P(needed no recrimping | passed inspection) = $\dfrac{P(passed.initially)}{P(passed.inspection)}$

$$= \frac{.95}{.974} = .9754$$

99. Let A = 1st functions, B = 2nd functions, so P(B) = .9, P(A ∪ B) = .96, P(A ∩ B)=.75.
Thus, P(A ∪ B) = P(A) + P(B) - P(A ∩ B) = P(A) + .9 - .75 = .96, implying P(A) = .81.
This gives P(B | A) = $\dfrac{P(B \cap A)}{P(A)} = \dfrac{.75}{.81} = .926$

101.

a. The law of total probability gives

$$P(late) = \sum_{i=1}^{3} P(late \mid E_i) \cdot P(E_i)$$

$$= (.02)(.40) + (.01)(.50) + (.05)(.10) = .018$$

b. $P(E_1' \mid \text{on time}) = 1 - P(E_1 \mid \text{on time})$

$$= 1 - \frac{P(E_1 \cap on.time)}{P(on.time)} = 1 - \frac{(.98)(.4)}{.982} = .601$$

103.

a. $P(\text{all different}) = \dfrac{(365)(364)...(356)}{(365)^{10}} = .883$

$P(\text{at least two the same}) = 1 - .883 = .117$

b. $P(\text{at least two the same}) = .476$ for k=22, and $= .507$ for k=23

c. $P(\text{at least two have the same SS number}) = 1 - P(\text{all different})$

$$= 1 - \frac{(1000)(999)...(991)}{(1000)^{10}}$$

$$= 1 - .956 = .044$$

Thus $P(\text{at least one "coincidence"}) = P(\text{BD coincidence} \cup \text{SS coincidence})$
$$= .117 + .044 - (.117)(.044) = .156$$

105. $P(\text{detection by the end of the nth glimpse}) = 1 - P(\text{not detected in } 1^{st} \text{ n})$
$= 1 - P(G_1' \cap G_2' \cap ... \cap G_n') = 1 - P(G_1')P(G_2') ... P(G_n')$

$$= 1 - (1 - p_1)(1 - p_2) ... (1 - p_n) = 1 - \prod_{i=1}^{n}(1 - p_i)$$

107.

a. P(all in correct room) = $\dfrac{1}{4 \times 3 \times 2 \times 1} = \dfrac{1}{24} = .0417$

b. The 9 outcomes which yield incorrect assignments are: 2143, 2341, 2413, 3142, 3412, 3421, 4123, 4321, and 4312, so P(all incorrect) = $\dfrac{9}{24} = .375$

109. Note: s = 0 means that the very first candidate interviewed is hired. Each entry below is the candidate hired for the given policy and outcome.

Outcome	s=0	s=1	s=2	s=3	Outcome	s=0	s=1	s=2	s=3
1234	1	4	4	4	3124	3	1	4	4
1243	1	3	3	3	3142	3	1	4	2
1324	1	4	4	4	3214	3	2	1	4
1342	1	2	2	2	3241	3	2	1	1
1423	1	3	3	3	3412	3	1	1	2
1432	1	2	2	2	3421	3	2	2	1
2134	2	1	4	4	4123	4	1	3	3
2143	2	1	3	3	4132	4	1	2	2
2314	2	1	1	4	4213	4	2	1	3
2341	2	1	1	1	4231	4	2	1	1
2413	2	1	1	3	4312	4	3	1	2
2431	2	1	1	1	4321	4	3	2	1

s	0	1	2	3
P(hire#1)	$\dfrac{6}{24}$	$\dfrac{11}{24}$	$\dfrac{10}{24}$	$\dfrac{6}{24}$

So s = 1 is best.

Chapter 2: Probability

111. $P(A_1) = P(\text{draw slip 1 or 4}) = \frac{1}{2}$; $P(A_2) = P(\text{draw slip 2 or 4}) = \frac{1}{2}$;

$P(A_3) = P(\text{draw slip 3 or 4}) = \frac{1}{2}$; $P(A_1 \cap A_2) = P(\text{draw slip 4}) = \frac{1}{4}$;

$P(A_2 \cap A_3) = P(\text{draw slip 4}) = \frac{1}{4}$; $P(A_1 \cap A_3) = P(\text{draw slip 4}) = \frac{1}{4}$

Hence $P(A_1 \cap A_2) = P(A_1)P(A_2) = \frac{1}{4}$, $P(A_2 \cap A_3) = P(A_2)P(A_3) = \frac{1}{4}$,

   $P(A_1 \cap A_3) = P(A_1)P(A_3) = \frac{1}{4}$, thus there exists pairwise independence

$P(A_1 \cap A_2 \cap A_3) = P(\text{draw slip 4}) = \frac{1}{4} \neq 1/8 = P(A_1)p(A_2)P(A_3)$, so the events are not mutually independent.

CHAPTER 3

Section 3.1

1.

S:	FFF	SFF	FSF	FFS	FSS	SFS	SSF	SSS
X:	0	1	1	1	2	2	2	3

3. M = the difference between the large and the smaller outcome with possible values 0, 1, 2, 3, 4, or 5; W = 1 if he sum of the two resulting numbers is even and W = 0 otherwise, a Bernoulli random variable.

5. No. In the experiment in which a coin is tossed repeatedly until a H results, let Y = 1 if the experiment terminates with at most 5 tosses and Y = 0 otherwise. The sample space is infinite, yet Y has only two possible values.

7.

  a. Possible values are 0, 1, 2, ..., 12; discrete

  b. With N = # on the list, values are 0, 1, 2, ... , N; discrete

  c. Possible values are 1, 2, 3, 4, ... ; discrete

  d. { x: $0 < x < \infty$ } if we assume that a rattlesnake can be arbitrarily short or long; not discrete

  e. With c = amount earned per book sold, possible values are 0, c, 2c, 3c, ... , 10,000c; discrete

  f. { y: $0 < y < 14$} since 0 is the smallest possible pH and 14 is the largest possible pH; not discrete

  g. With m and M denoting the minimum and maximum possible tension, respectively, possible values are { x: $m < x < M$ }; not discrete

  h. Possible values are 3, 6, 9, 12, 15, ... -- i.e. 3(1), 3(2), 3(3), 3(4), ...giving a first element, etc,; discrete

9.

    a.   Returns to 0 can occur only after an even number of tosses; possible S values are 2, 4, 6, 8, …(i.e. 2(1), 2(2), 2(3), 2(4),…) an infinite sequence, so x is discrete.

    b.   Now a return to 0 is possible after any number of tosses greater than 1, so possible values are 2, 3, 4, 5, … (1+1,1+2, 1+3, 1+4, …, an infinite sequence) and X is discrete

Section 3.2

11.

    a.

x	4	6	8
P(x)	.45	.40	.15

    b.

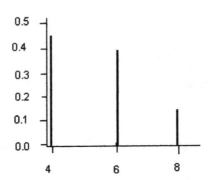

    c. $P(x \geq 6) = .40 + .15 = .55$ $P(x > 6) = .15$

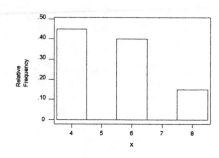

13.

    a. $P(X \leq 3) = p(0) + p(1) + p(2) + p(3) = .10 + .15 + .20 + .25 = .70$

    b. $P(X < 3) = P(X \leq 2) = p(0) + p(1) + p(2) = .45$

    c. $P(3 \leq X) = p(3) + p(4) + p(5) + p(6) = .55$

    d. $P(2 \leq X \leq 5) = p(2) + p(3) + p(4) + p(5) = .71$

    e.   The number of lines not in use is $6 - X$, so $6 - X = 2$ is equivalent to $X = 4$, $6 - X = 3$ to $X = 3$, and $6 - X = 4$ to $X = 2$. Thus we desire $P(2 \leq X \leq 4) = p(2) + p(3) + p(4) = .65$

    f. $6 - X \geq 4$ if $6 - 4 \geq X$, i.e. $2 \geq X$, or $X \leq 2$, and $P(X \leq 2) = .10 + .15 + .20 = .45$

Chapter 3: Discrete Random Variables and Probability Distributions

15.

a. (1,2) (1,3) (1,4) (1,5) (2,3) (2,4) (2,5) (3,4) (3,5) (4,5)

b. $P(X = 0) = p(0) = P[\{ (3,4) (3,5) (4,5)\}] = \frac{3}{10} = .3$

$P(X = 2) = p(2) = P[\{ (1,2) \}] = \frac{1}{10} = .1$

$P(X = 1) = p(1) = 1 - [p(0) + p(2)] = .60$, and $p(x) = 0$ if $x \neq 0, 1, 2$

c. $F(0) = P(X \leq 0) = P(X = 0) = .30$

$F(1) = P(X \leq 1) = P(X = 0 \text{ or } 1) = .90$

$F(2) = P(X \leq 2) = 1$

The c.d.f. is

$$F(x) = \begin{cases} 0 & x < 0 \\ .30 & 0 \leq x < 1 \\ .90 & 1 \leq x < 2 \\ 1 & 2 \leq x \end{cases}$$

17.

a. $P(2) = P(Y = 2) = P(1^{st} \text{ 2 batteries are acceptable})$

$\qquad = P(AA) = (.9)(.9) = .81$

b. $p(3) = P(Y = 3) = P(UAA \text{ or } AUA) = (.1)(.9)^2 + (.1)(.9)^2 = 2[(.1)(.9)^2] = .162$

c. The fifth battery must be an A, and one of the first four must also be an A. Thus, $p(5) = P(AUUUA \text{ or } UAUUA \text{ or } UUAUA \text{ or } UUUAA) = 4[(.1)^3(.9)^2] = .00324$

d. $P(Y = y) = p(y) = P(\text{the } y^{th} \text{ is an A and so is exactly one of the first } y - 1)$

$\qquad = (y - 1)(.1)^{y-2}(.9)^2, \ y = 2,3,4,5,...$

19. Let A denote the type O+ individual (type O positive blood) and B, C, D, the other 3 individuals. Then $p(1) - P(Y = 1) = P(A \text{ first}) = \frac{1}{4} = .25$

$p(2) = P(Y = 2) = P(B, C, \text{ or } D \text{ first and A next}) = \frac{3}{4} \cdot \frac{1}{3} = \frac{1}{4} = .25$

$p(4) = P(Y = 3) = P(A \text{ last}) = \frac{3}{4} \cdot \frac{2}{3} \cdot \frac{1}{2} = \frac{1}{4} = .25$

So $p(3) = 1 - (.25 + .25 + .25) = .25$

Chapter 3: Discrete Random Variables and Probability Distributions

21. The jumps in F(x) occur at x = 0, 1, 2, 3, 4, 5, and 6, so we first calculate F( ) at each of these values:

$$F(0) = P(X \leq 0) = P(X = 0) = .10$$
$$F(1) = P(X \leq 1) = p(0) + p(1) = .25$$
$$F(2) = P(X \leq 2) = p(0) + p(1) + p(2) = .45$$
$$F(3) = .70, F(4) = .90, F(5) = .96, \text{ and } F(6) = 1.$$

The c.d.f. is

$$F(x) = \begin{cases} .00 & x < 0 \\ .10 & 0 \leq x < 1 \\ .25 & 1 \leq x < 2 \\ .45 & 2 \leq x < 3 \\ .70 & 3 \leq x < 4 \\ .90 & 4 \leq x < 5 \\ .96 & 5 \leq x < 6 \\ 1.00 & 6 \leq x \end{cases}$$

Then $P(X \leq 3) = F(3) = .70$, $P(X < 3) = P(X \leq 2) = F(2) = .45$,
$P(3 \leq X) = 1 - P(X \leq 2) = 1 - F(2) = 1 - .45 = .55$,
and $P(2 \leq X \leq 5) = F(5) - F(1) = .96 - .25 = .71$

23.

a. Possible X values are those values at which F(x) jumps, and the probability of any particular value is the size of the jump at that value. Thus we have:

x	1	3	4	6	12
p(x)	.30	.10	.05	.15	.40

b. $P(3 \leq X \leq 6) = F(6) - F(3-) = .60 - .30 = .30$
$P(4 \leq X) = 1 - P(X < 4) = 1 - F(4-) = 1 - .40 = .60$

25.

a. Possible X values are 1, 2, 3, ...

$$P(1) = P(X = 1) = P(\text{return home after just one visit}) = \tfrac{1}{3}$$

$$P(2) = P(X = 2) = P(\text{second visit and then return home}) = \tfrac{2}{3} \cdot \tfrac{1}{3}$$

$$P(3) = P(X = 3) = P(\text{three visits and then return home}) = \left(\tfrac{2}{3}\right)^2 \cdot \tfrac{1}{3}$$

In general $p(x) = \left(\tfrac{2}{3}\right)^{x-1}\left(\tfrac{1}{3}\right)$ for $x = 1, 2, 3, \dots$

b. The number of straight line segments is $Y = 1 + X$ (since the last segment traversed returns Alvie to O), so as in a, $p(y) = \left(\tfrac{2}{3}\right)^{y-2}\left(\tfrac{1}{3}\right)$ for $y = 2, 3, \dots$

c. Possible Z values are 0, 1, 2, 3, ...

$p(0) = P(\text{male first and then home}) = \tfrac{1}{2} \cdot \tfrac{1}{3} = \tfrac{1}{6},$

$p(1) = P(\text{exactly one visit to a female}) = P(\text{female 1}^{\text{st}}, \text{then home}) + P(F, M,$
home$) + P(M, F, \text{home}) + P(M, F, M, \text{home})$

$= \left(\tfrac{1}{2}\right)\left(\tfrac{1}{3}\right) + \left(\tfrac{1}{2}\right)\left(\tfrac{2}{3}\right)\left(\tfrac{1}{3}\right) + \left(\tfrac{1}{2}\right)\left(\tfrac{2}{3}\right)\left(\tfrac{1}{3}\right) + \left(\tfrac{1}{2}\right)\left(\tfrac{2}{3}\right)\left(\tfrac{2}{3}\right)\left(\tfrac{1}{3}\right)$

$= \left(\tfrac{1}{2}\right)\left(1 + \tfrac{2}{3}\right)\left(\tfrac{1}{3}\right) + \left(\tfrac{1}{2}\right)\left(\tfrac{2}{3}\right)\left(\tfrac{2}{3} + 1\right)\left(\tfrac{1}{3}\right) = \left(\tfrac{1}{2}\right)\left(\tfrac{5}{3}\right)\left(\tfrac{1}{3}\right) + \left(\tfrac{1}{2}\right)\left(\tfrac{2}{3}\right)\left(\tfrac{5}{3}\right)\left(\tfrac{1}{3}\right)$

where the first term corresponds to initially visiting a female and the second term corresponds to initially visiting a male. Similarly,

$p(2) = \left(\tfrac{1}{2}\right)\left(\tfrac{2}{3}\right)^2\left(\tfrac{5}{3}\right)\left(\tfrac{1}{3}\right) + \left(\tfrac{1}{2}\right)\left(\tfrac{2}{3}\right)^2\left(\tfrac{5}{3}\right)\left(\tfrac{1}{3}\right).$ In general,

$p(z) = \left(\tfrac{1}{2}\right)\left(\tfrac{2}{3}\right)^{2z-2}\left(\tfrac{5}{3}\right)\left(\tfrac{1}{3}\right) + \left(\tfrac{1}{2}\right)\left(\tfrac{2}{3}\right)^{2z-2}\left(\tfrac{5}{3}\right)\left(\tfrac{1}{3}\right) = \left(\tfrac{24}{54}\right)\left(\tfrac{2}{3}\right)^{2z-2}$ for $z = 1, 2, 3, \dots$

27. If $x_1 < x_2$, $F(x_2) = P(X \le x_2) = P(\{X \le x_1\} \cup \{x_1 < X \le x_2\})$
$\qquad\qquad = P(X \le x_1) + P(x_1 < X \le x_2) \ge P(X \le x_1) = F(x_1).$
$F(x_1) = F(x_2)$ when $P(x_1 < X \le x_2) = 0.$

Section 3.3

29.

a. $E(Y) = \sum_{x=0}^{4} y \cdot p(y) = (0)(.60) + (1)(.25) + (2)(.10) + (3)(.05) = .60$

b. $E(100Y^2) = \sum_{x=0}^{4} 100y^2 \cdot p(y) = (0)(.60) + (100)(.25)$

$$+ (400)(.10) + (900)(.05) = 110$$

31.

a. $E(X) = (13.5)(.2) + (15.9)(.5) + (19.1)(.3) = 16.38,$
$E(X^2) = (13.5)^2(.2) + (15.9)^2(.5) + (19.1)^2(.3) = 272.298,$
$V(X) = 272.298 - (16.38)^2 = 3.9936$

b. $E(25X - 8.5) = 25 E(X) - 8.5 = (25)(16.38) - 8.5 = 401$

c. $V(25X - 8.5) = V(25X) = (25)^2 V(X) = (625)(3.9936) = 2496$

d. $E[h(X)] = E[X - .01X^2] = E(X) - .01E(X^2) = 16.38 - 2.72 = 13.66$

33. $E(X) = \sum_{x=1}^{\infty} x \cdot p(x) = \sum_{x=1}^{\infty} x \cdot \frac{c}{x^3} = c \sum_{x=1}^{\infty} \frac{1}{x^2}$, but it is a well-known result from the

theory of infinite series that $\sum_{x=1}^{\infty} \frac{1}{x^2} < \infty$, so E(X) is finite.

35.

P(x)	.8	.1	.08	.02
x	0	1,000	5,000	10,000
H(x)	0	500	4,500	9,500

$E[h(X)] = 600$. Premium should be $100 plus expected value of damage minus deductible or $700.

37. $E[h(X)] = E\left(\dfrac{1}{X}\right) = \sum\limits_{x=1}^{6}\left(\dfrac{1}{x}\right)\cdot p(x) = \dfrac{1}{6}\sum\limits_{x=1}^{6}\dfrac{1}{x} = .408$, whereas $\dfrac{1}{3.5} = .286$, so
you expect to win more if you gamble.

39.

   a. The line graph of the p.m.f. of $-X$ is just the line graph of the p.m.f. of X
reflected about zero, but both have the same degree of spread about their
respective means, suggesting $V(-X) = V(X)$.

   b. With $a = -1$, $b = 0$, $V(aX + b) = V(-X) = a^2 V(X)$.

41.

   a. $E[X(X-1)] = E(X^2) - E(X)$, $\Rightarrow E(X^2) = E[X(X-1)] + E(X) = 32.5$

   b. $V(X) = 32.5 - (5)^2 = 7.5$

   c. $V(X) = E[X(X-1)] + E(X) - [E(X)]^2$

43.

   a.

k	2	3	4	5	10
$\dfrac{1}{k^2}$	.25	.11	.06	.04	.01

   b. $\mu = \sum\limits_{x=0}^{6} x\cdot p(x) = 2.64$,

$$\sigma^2 = \left[\sum\limits_{x=0}^{6} x^2\cdot p(x)\right] - \mu^2 = 2.37,\ \sigma = 1.54$$

Thus $\mu - 2\sigma = -.44$, and $\mu + 2\sigma = 5.72$,
so $P(|x-\mu| \geq 2\sigma) = P(X$ is lat least 2 s.d.'s from $\mu)$
                   $= P(x$ is either $\leq -.44$ or $\geq 5.72) = P(X = 6) = .04$.
Chebyshev's bound of .025 is much too conservative. For K = 3,4,5, and 10,
$P(|x-\mu| \geq k\sigma) = 0$, here again pointing to the very conservative nature of the
bound $\dfrac{1}{k^2}$.

    c. $\mu = 0$ and $\sigma = \frac{1}{3}$, so $P(|x\text{-}\mu| \geq 3\sigma) = P(|\,X\,| \geq 1)$

                               $= P(X = \text{-}1 \text{ or } +1) = \frac{1}{18} + \frac{1}{18} = \frac{1}{9}$, identical to the upper bound.

    d.   Let $p(\text{-}1) = \frac{1}{50}, p(+1) = \frac{1}{50}, p(0) = \frac{24}{25}$.

Section 3.4

45.

    a. $B(4;10,.3) = .850$

    b. $b(4;10,.3) = B(4;10,.3) - B(3;10,.3) = .200$

    c. $b(6;10,.7) = B(6;10,.7) - B(5;10,.7) = .200$

    d. $P(\,2 \leq X \leq 4) = B(4;10,.3) - B(1;10,.3) = .701$

    e. $P(2 < X) = 1 - P(X \leq 1) = 1 - B(1;10,.3) = .851$

    f. $P(X \leq 1) = B(1;10,.7) = .0000$

    g. $P(2 < X < 6) = P(\,3 \leq X \leq 5) = B(5;10,.3) - B(2;10,.3) = .570$

47. $X \sim \text{Bin}(6, .10)$

a. $P(X = 1) = \binom{n}{x}(p)^x(1-p)^{n-x} = \binom{6}{1}(.1)^1(.9)^5 = .3543$

b. $P(X \geq 2) = 1 - [P(X = 0) + P(X = 1)]$.

From **a** , we know $P(X = 1) = .3543$, and $P(X = 0) = \binom{6}{0}(.1)^0(.9)^6 = .5314$.

Hence $P(X \geq 2) = 1 - [.3543 + .5314] = .1143$

c. Either 4 or 5 goblets must be selected

i)   Select 4 goblets with zero defects: $P(X = 0) = \binom{4}{0}(.1)^0(.9)^4 = .6561$.

ii) Select 4 goblets, one of which has a defect, and the 5th is good:

$$\left[\binom{4}{1}(.1)^1(.9)^3\right] \times .9 = .26244$$

So the desired probability is $.6561 + .26244 = .91854$

49.   Let $S = $ has at least one citation. Then $p = .4$, $n = 15$

a. If at least 10 have no citations (Failure), then at most 5 have had at least one (Success): $P(X \leq 5) = B(5;15,.40) = .403$

b. $P(X \leq 7) = B(7;15,.40) = .787$

c. $P(5 \leq X \leq 10) = P(X \leq 10) - P(X \leq 4) = .991 - .217 = .774$

51.   Let S represent a telephone that is submitted for service while under warranty and must be replaced. Then $p = P(S) = P(\text{replaced} \mid \text{submitted}) \cdot P(\text{submitted}) = (.40)(.20) = .08$. Thus X, the number among the company's 10 phones that must be replaced, has a binomial distribution with $n = 10$, $p = .08$, so $p(2) = P(X=2) = $

$$\binom{10}{2}(.08)^2(.92)^8 = .1478$$

53.   X = the number of flashlights that work.
Let event B = {battery has acceptable voltage}.
Then P(flashlight works) = P(both batteries work) = P(B)P(B) = (.9)(.9) = .81 We
must assume that the batteries' voltage levels are independent.
X~ Bin (10, .81). P(X≥9) = P(X=9) + P(X=10)

$$\binom{10}{9}(.81)^9(.19)+\binom{10}{10}(.81)^{10} = .285 + .122 = .407$$

55.

   a. P(rejecting claim when p = .8) = B(15;25,.8) = .017

   b. P(not rejecting claim when p = .7) = P(X ≥ 16 when p = .7)
= 1 - B(15;25,.7) = 1 - .189 = .811; for p = .6, this probability is
= 1 - B(15;25,.6) = 1 - .575 = .425.

   c. The probability of rejecting the claim when p = .8 becomes B(14;25,.8) = .006,
smaller than in **a** above. However, the probabilities of **b** above increase to .902
and .586, respectively.

57.   If topic A is chosen, when n = 2, P(at least half received)
= P(X ≥ 1) = 1 – P(X = 0) = 1 – (.1)² = .99
If B is chosen, when n = 4, P(at least half received)
= P(X ≥ 2) = 1 – P(X ≤ 1) = 1 – (0.1)⁴ – 4(.1)³(.9) = .9963
Thus topic B should be chosen.
If p = .5, the probabilities are .75 for A and .6875 for B, so now A should be chosen.

59.

a. $b(x; n, 1-p) = \binom{n}{x}(1-p)^x(p)^{n-x} = \binom{n}{n-x}(p)^{n-x}(1-p)^x = b(n-x; n, p)$

Alternatively, $P(x$ S's when $P(S) = 1 - p) = P(n-x$ F's when $P(F) = p)$, since the two events are identical), but the labels S and F are arbitrary so can be interchanged (if $P(S)$ and $P(F)$ are also interchanged), yielding $P(n-x$ S's when $P(S) = 1 - p)$ as desired.

b. $B(x;n,1-p) = P(\text{at most } x \text{ S's when } P(S) = 1-p)$
$= P(\text{at least } n-x \text{ F's when } P(F) = p)$
$= P(\text{at least } n-x \text{ S's when } P(S) = p)$
$= 1 - P(\text{at most } n-x-1 \text{ S's when } P(S) = p)$
$= 1 - B(n-x-1;n,p)$

c. Whenever $p > .5$, $(1-p) < .5$ so probabilities involving X can be calculated using the results **a** and **b** in combination with tables giving probabilities only for $p \leq .5$

61.

a. Although there are three payment methods, we are only concerned with S = uses a debit card and F = does not use a debit card. Thus we can use the binomial distribution. So $n = 100$ and $p = .5$. $E(X) = np = 100(.5) = 50$, and $V(X) = 25$.

b. With S = doesn't pay with cash, $n = 100$ and $p = .7$, $E(X) = np = 100(.7) = 70$, and $V(X) = 21$.

63. When $p = .5$, $\mu = 10$ and $\sigma = 2.236$, so $2\sigma = 4.472$ and $3\sigma = 6.708$. The inequality $|X - 10| \geq 4.472$ is satisfied if either $X \leq 5$ or $X \geq 15$, or $P(|X - \mu| \geq 2\sigma)$ $= P(X \leq 5 \text{ or } X \geq 15) = .021 + .021 = .042$.

In the case $p = .75$, $\mu = 15$ and $\sigma = 1.937$, so $2\sigma = 3.874$ and $3\sigma = 5.811$. $P(|X - 15| \geq 3.874) = P(X \leq 11 \text{ or } X \geq 19) = .041 + .024 = .065$, whereas $P(|X - 15| \geq 5.811) = P(X \leq 9) = .004$. All these probabilities are considerably less than the upper bounds .25(for $k = 2$) and .11 (for $k = 3$) given by Chebyshev.

Section 3.5

65.   X~h(x; 6, 12, 7)

a. $P(X=5) = \dfrac{\dbinom{7}{5}\dbinom{5}{1}}{\dbinom{12}{6}} = \dfrac{105}{924} = .114$

b. $P(X\leq4) = 1 - P(X\geq5) = 1 - [P(X=5) + P(X=6)] =$

$$1 - \left[\frac{\dbinom{7}{5}\dbinom{5}{1}}{\dbinom{12}{6}} + \frac{\dbinom{7}{6}}{\dbinom{12}{6}}\right] = 1 - \frac{105 + 7}{924} = 1 - .121 = .879$$

c. $E(X) = \left(\dfrac{6 \cdot 7}{12}\right) = 3.5$; $\sigma = \sqrt{\left(\frac{6}{11}\right)\left(6\right)\left(\frac{7}{12}\right)\left(\frac{5}{12}\right)} = \sqrt{.795} = .892$

$P(X > 3.5 + .892) = P(X > 4.392) = P(X \geq 5) = .121$ (see part b)

d.   We can approximate the hypergeometric distribution with the binomial if the population size and the number of successes are large: h(x;15,40,400) approaches b(x;15,.10). So $P(X\leq5) \approx B(5; 15, .10)$ from the binomial tables = .998

67.

a. Possible values of X are 5, 6, 7, 8, 9, 10. (In order to have less than 5 of the granite, there would have to be more than 10 of the basaltic).

$$P(X = 5) = h(5;\ 15,10,20) = \frac{\binom{10}{5}\binom{10}{10}}{\binom{20}{15}} = .0163 .$$

Following the same pattern for the other values, we arrive at the pmf, in table form below.

x	5	6	7	8	9	10
p(x)	.0163	.1354	.3483	.3483	.1354	.0163

b. P(all 10 of one kind or the other) = P(X = 5) + P(X = 10) = .0163 + .0163 = .0326

c. $E(X) = n \cdot \dfrac{M}{N} = 15 \cdot \dfrac{10}{20} = 7.5$; $V(X) = \left(\dfrac{5}{19}\right)(7.5)\left(1 - \dfrac{10}{20}\right) = .9868$;

$\sigma_x = .9934$

$\mu \pm \sigma = 7.5 \pm .9934 = (6.5066, 8.4934)$, so we want
$P(X = 7) + P(X = 8) = .3483 + .3483 = .6966$

69.

a. h(x; 10,10,20) (the successes here are the top 10 pairs, and a sample of 10 pairs is drawn from among the 20)

b. Let X = the number among the top 5 who play E-W. Then P(all of top 5 play the same direction) = P(X = 5) + P(X = 0) = h(5;10,5,20) + h(5;10,5,20)

$$= \frac{\binom{15}{5}}{\binom{20}{10}} + \frac{\binom{15}{10}}{\binom{20}{10}} = .033$$

c. N = 2n; M = n; n = n
    h(x;n,n,2n)

$$E(X) = n \cdot \frac{n}{2n} = \frac{1}{2}n;$$

$$V(X) =$$

$$\left(\frac{2n-n}{2n-1}\right) \cdot n \cdot \frac{n}{2n} \cdot \left(1 - \frac{n}{2n}\right) = \left(\frac{n}{2n-1}\right) \cdot \frac{n}{2} \cdot \left(1 - \frac{n}{2n}\right) = \left(\frac{n}{2n-1}\right) \cdot \frac{n}{2} \cdot \left(\frac{1}{2}\right)$$

71.

a. With S = a female child and F = a male child, let X = the number of F's before the 2nd S. Then P(X = x) = nb(x;2, .5)

b. P(exactly 4 children) = P(exactly 2 males)
    $$= nb(2;2,.5) = (3)(.0625) = .188$$

c. P(at most 4 children) = P(X ≤ 2)

$$= \sum_{x=0}^{2} nb(x;2,.5) = .25 + 2(.25)(.5) + 3(.0625) = .688$$

d. $E(X) = \dfrac{(2)(.5)}{.5} = 2$, so the expected number of children = E(X + 2)
    $$= E(X) + 2 = 4$$

73. This is identical to an experiment in which a single family has children until exactly 6 females have been born(since p = .5 for each of the three families), so p(x) = nb(x;6,.5) and E(X) = 6 (= 2+2+2, the sum of the expected number of males born to each one.)

Section 3.6

75.

    a. $P(X \leq 8) = F(8;5) = .932$

    b. $P(X = 8) = F(8;5) - F(7;5) = .065$

    c. $P(X \geq 9) = 1 - P(X \leq 8) = .068$

    d. $P(5 \leq X \leq 8) = F(8;5) - F(4;5) = .492$

77.

    a. $P(X \leq 10) = F(10;20) = .011$

    b. $P(X > 20) = 1 - F(20;20) = 1 - .559 = .441$

    c. $P(10 \leq X \leq 20) = F(20;20) - F(9;20) = .559 - .005 = .554$
           $P(10 < X < 20) = F(19;20) - F(10;20) = .470 - .011 = .459$

    d. $E(X) = \lambda = 20, \ \sigma_X = \sqrt{\lambda} = 4.472$
        $P(\mu - 2\sigma < X < \mu + 2\sigma)$ $= P(20 - 8.944 < X < 20 + 8.944)$
                                        $= P(11.056 < X < 28.944)$
                                        $= P(X \leq 28) - P(X \leq 11)$
                                        $= F(28;20) - F(12;20)]$
                                        $= .966 - .021 = .945$

79. $p = \dfrac{1}{200}; n = 1000; \lambda = np = 5$

    a. $P(5 \leq X \leq 8) = F(8;5) - F(4;5) = .492$

    b. $P(X \geq 8) = 1 - P(X \leq 7) = 1 - .867 = .133$

81.

    a. $\lambda = 8$ when $t = 1$, so $P(X = 6) = F(6;8) - F(5;8) = .313 - .191 = .122$,
               $P(X \geq 6) = 1 - F(5;8) = .809$, and $P(X \geq 10) = 1 - F(9;8) = .283$

    b. $t = 90$ min $= 1.5$ hours, so $\lambda = 12$; thus the expected number of arrivals is 12 and
       the SD $= \sqrt{12} = 3.464$

    c. $t = 2.5$ hours implies that $\lambda = 20$; in this case, $P(X \geq 20) = 1 - F(19;20) = .530$
       and $P(X \leq 10) = F(10;20) = .011$.

83.

    a. For a two hour period the parameter of the distribution is $\lambda t = (4)(2) = 8$,
       so $P(X = 10) = F(10;8) - F(9;8) = .099$.

    b. For a 30 minute period, $\lambda t = (4)(.5) = 2$, so $P(X = 0) = F(0,2) = .135$

    c. $E(X) = \lambda t = 2$

85. $= 1/(\text{mean time between occurrences}) = \dfrac{1}{.5} = 2$

    a. $\alpha t = (2)(2) = 4$

    b. $P(X > 5) 1 - P(X \leq 5) = 1 - .785 = .215$

    c. Solve for t, given $\alpha = 2$:
         $.1 = e^{-\alpha t}$
         $\ln(.1) = -\alpha t$
         $t = \dfrac{2.3026}{2} \approx 1.15$ years

87.

    a. For a one-quarter acre plot, the parameter is $(80)(.25) = 20$,
       so $P(X \leq 16) = F(16;20) = .221$

    b. The expected number of trees is $\lambda \cdot (\text{area}) = 80(85,000) = 6,800,000$.

    c. The area of the circle is $\pi r^2 = .031416$ sq. miles or 20.106 acres. Thus X has a
       Poisson distribution with parameter 20.106

89.

    a. No events in $(0, t+\Delta t)$ if and only if no events in (o, t) and no events in $(t, t+\Delta t)$.
Thus, $P_0 (t+\Delta t) = P_0(t) \cdot P(\text{no events in } (t, t+\Delta t))$
$= P_0(t)[1 - \lambda \cdot \Delta t - o(\Delta t)]$

    b. $\dfrac{P_0(t + \Delta t) - P_0(t)}{\Delta t} = -\lambda P_0(t)\dfrac{\Delta' t}{\Delta' t} - P_0(t) \cdot \dfrac{o(\Delta t)}{\Delta t}$

    c. $\dfrac{d}{dt}\left[e^{-\lambda t}\right] = -\lambda e^{-\lambda t} = -\lambda P_0(t)$, as desired.

    d. $\dfrac{d}{dt}\left[\dfrac{e^{-\lambda t}(\lambda t)^k}{k!}\right] = \dfrac{-\lambda e^{-\lambda t}(\lambda t)^k}{k!} + \dfrac{k\lambda e^{-\lambda t}(\lambda t)^{k-1}}{k!}$

        $= -\lambda\dfrac{e^{-\lambda t}(\lambda t)^k}{k!} + \lambda\dfrac{e^{-\lambda t}(\lambda t)^{k-1}}{(k-1)!} = -\lambda P_k(t) + \lambda P_{k-1}(t)$ as desired.

Supplementary Exercises

91.

a. p(1) = P(exactly one suit) = P(all spades) + P(all hearts) + P(all diamonds)

$$+ \text{ P(all clubs)} = 4\text{P(all spades)} = 4 \cdot \frac{\dbinom{13}{5}}{\dbinom{52}{5}} = .00198$$

p(2) = P(all hearts and spades with at least one of each) + ...+ P(all diamonds and clubs with at least one of each)

= 6 P(all hearts and spades with at least one of each)

= 6 [P(1 h and 4 s) + P(2 h and 3 s) + P(3 h and 2 s) + P(4 h and 1 s)]

$$= 6 \cdot \left[2 \cdot \frac{\dbinom{13}{4}\dbinom{13}{1}}{\dbinom{52}{5}} + 2 \cdot \frac{\dbinom{13}{3}\dbinom{13}{2}}{\dbinom{52}{5}} \right] = 6 \left[\frac{18,590 + 44,616}{2,598,960} \right] = .14592$$

$$\text{p(4) = 4P(2 spades, 1 h, 1 d, 1 c)} = \frac{4 \cdot \dbinom{13}{2}(13)(13)(13)}{\dbinom{52}{5}} = .26375$$

p(3) = 1 − [p(1) + p(2) + p(4)] = .58835

b. μ =

$$\sum_{x=1}^{4} x \cdot p(x) = 3.114, \ \sigma^2 = \left[\sum_{x=1}^{4} x^2 \cdot p(x) \right] - (3.114)^2 = .405, \sigma = .636$$

93.

    a. b(x;15,.75)

    b. $P(X > 10) \doteq 1 - B(9;15, .75) = 1 - .148$

    c. $B(10;15, .75) - B(5;15, .75) = .314 - .001 = .313$

    d. $\mu = (15)(.75) = 11.75, \sigma^2 = (15)(.75)(.25) = 2.81$

    e. Requests can all be met if and only if $X \leq 10$, and $15 - X \leq 8$, i.e. if $7 \leq X \leq 10$, so P(all requests met) $= B(10; 15,.75) - B(6; 15,.75) = .310$

95.   Let $X \sim Bin(5, .9)$. Then $P(X \geq 3) = 1 - P(X \leq 2) = 1 - B(2;5,.9) = .991$

97.

    a. $N = 500, p = .005$, so $np = 2.5$ and $b(x; 500, .005) \doteq p(x; 2.5)$, a Poisson p.m.f.

    b. $P(X = 5) = p(5; 2.5) - p(4; 2.5) = .9580 - .8912 = .0668$

    c. $P(X \geq 5) = 1 - p(4;2.5) = 1 - .8912 = .1088$

99.   Let Y denote the number of tests carried out. For $n = 3$, possible Y values are 1 and 4. $P(Y = 1) = P(\text{no one has the disease}) = (.9)^3 = .729$ and $P(Y = 4) = .271$, so $E(Y) = (1)(.729) + (4)(.271) = 1.813$, as contrasted with the 3 tests necessary without group testing.

101. $p(2) = P(X = 2) = P(\text{S on \#1 and S on \#2}) = p^2$
        $p(3) = P(\text{S on \#3 and S on \#2 and F on \#1}) = (1 - p)p^2$
        $p(4) = P(\text{S on \#4 and S on \#3 and F on \#2}) = (1 - p)p^2$
        $p(5) = P(\text{S on \#5 and S on \#4 and F on \#3 and no 2 consecutive S's on trials prior to}$
        \#3$) = [1 - p(2)](1 - p)p^2$
        $p(6) = P(\text{S on \#6 and S on \#5 and F on \#4 and no 2 consecutive S's on trials prior to}$
        \#4$) = [1 - p(2) - p(3)](1 - p)p^2$
        In general, for $x = 5, 6, 7, ...$: $p(x) = [1 - p(2) - ... - p(x - 3)](1 - p)p^2$
        For $p = .9$,

x	2	3	4	5	6	7	8
p(x)	.81	.081	.081	.0154	.0088	.0023	.0010

So $P(X \leq 8) = p(2) + ... + p(8) = .9995$

Chapter 3: Discrete Random Variables and Probability Distributions

103.

   a. Let event C = seed carries single spikelets, and event P = seed produces ears with single spikelets. Then $P(P \cap C) = P(P \mid C) \cdot P(C) = .29 (.40) = .116$. Let X = the number of seeds out of the 10 selected that meet the condition $P \cap C$. Then X ~ Bin(10, .116). $P(X = 5) = \binom{10}{5}(.116)^5(.884)^5 = .002857$

   b. For 1 seed, the event of interest is P = seed produces ears with single spikelets.
$P(P) = P(P \cap C) + P(P \cap C') = .116$ (from **a**) $+ P(P \mid C') \cdot P(C')$
$= .116 + (.26)(.40) = .272$.
Let Y = the number out of the 10 seeds that meet condition P.
Then Y ~ Bin(10, .272), and $P(Y = 5) = .0767$.
$P(Y \le 5) = b(0;10,.272) + ... + b(5;10,.272) = .041813 + ... + .076719 = .97024$

105.

   a. $P(X = 0) = F(0;2)\ 0.135$

   b. Let S = an operator who receives no requests. Then p = .135 and we wish P(4 S's in 5 trials) $= b(4;5,..135) = \binom{5}{4}(.135)^4(.884)^1 = .00144$

   c. P(all receive x) = P(first receives x) $\cdot$ ... $\cdot$ P(fifth receives x) $= \left[\dfrac{e^{-2}2^x}{x!}\right]^5$, and

   P(all receive the same number) is the sum from x = 0 to ∞.

107. The number sold is min (X, 5), so $E[\min(x, 5)] = \sum_{}^{\infty} \min(x,5)p(x;4)$

$= (0)p(0;4) + (1)\ p(1;4) + (2)\ p(2;4) + (3)\ p(3;4) + (4)\ p(4;4) + 5\sum_{x=5}^{\infty} p(x;4)$

$= 1.735 + 5[1 - F(4;4)] = 3.59$

109.

    a. No; probability of success is not the same for all tests

    b. There are four ways exactly three could have positive results. Let D represent those with the disease and D′ represent those without the disease.

Combination		**Probability**
D	**D′**	
0	3	$\left[\binom{5}{0}(.2)^{0}(.8)^{5}\right]\cdot\left[\binom{5}{3}(.9)^{3}(.1)^{2}\right]$
		$=(.32768)(.0729)=.02389$
1	2	$\left[\binom{5}{1}(.2)^{1}(.8)^{4}\right]\cdot\left[\binom{5}{2}(.9)2(.1)^{3}\right]$
		$=(.4096)(.0081)=.00332$
2	1	$\left[\binom{5}{2}(.2)^{2}(.8)^{3}\right]\cdot\left[\binom{5}{1}(.9)^{1}(.1)^{4}\right]$
		$=(.2048)(.00045)=.00009216$
3	0	$\left[\binom{5}{3}(.2)^{3}(.8)^{2}\right]\cdot\left[\binom{5}{0}(.9)^{0}(.1)^{5}\right]$
		$=(.0512)(.00001)=.000000512$

Adding up the probabilities associated with the four combinations yields 0.0273.

111.

a. $p(x;\lambda,\mu) = \frac{1}{2}p(x;\lambda) + \frac{1}{2}p(x;\mu)$ where both $p(x;\lambda)$ and $p(x;\mu)$ are Poisson p.m.f.'s and thus ≥ 0, so $p(x;\lambda,\mu) \geq 0$. Further,

$$\sum_{x=0}^{\infty} p(x;\lambda,\mu) = \frac{1}{2}\sum_{x=0}^{\infty} p(x;\lambda) + \frac{1}{2}\sum_{x=0}^{\infty} p(x;\mu) = \frac{1}{2} + \frac{1}{2} = 1$$

b. $.6p(x;\lambda) + .4p(x;\mu)$

c. $E(X) = \sum_{x=0}^{\infty} x[\frac{1}{2}p(x;\lambda) + \frac{1}{2}p(x;\mu)] = \frac{1}{2}\sum_{x=0}^{\infty} xp(x;\lambda) + \frac{1}{2}\sum_{x=0}^{\infty} xp(x;\mu)$

$= \frac{1}{2}\lambda + \frac{1}{2}\mu = \frac{\lambda+\mu}{2}$

d. $E(X^2) =$

$\frac{1}{2}\sum_{x=0}^{\infty} x^2 p(x;\lambda) + \frac{1}{2}\sum_{x=0}^{\infty} x^2 p(x;\mu) = \frac{1}{2}(\lambda^2 + \lambda) + \frac{1}{2}(\mu^2 + \mu)$ (since for a

Poisson r.v., $E(X^2) = V(X) + [E(X)]^2 = \lambda + \lambda^2$),

so $V(X) = \frac{1}{2}[\lambda^2 + \lambda + \mu^2 + \mu] - [\frac{\lambda+\mu}{2}]^2 = (\frac{\lambda-\mu}{2})^2 + \frac{\lambda+\mu}{2}$

113. $P(X=j) = \sum_{i=1}^{10} P \text{ (arm on track } i \cap X=j) = \sum_{i=1}^{10} P\text{ (}X=j\mid \text{arm on } i \text{)} \cdot p_i$

$= \sum_{i=1}^{10} P \text{ (next seek at I+j+1 or I-j-1)} \cdot p_i = \sum_{i=1}^{10} (p_{i+j+1} + P_{i-j-1})p_i$

where $p_k = 0$ if $k < 0$ or $k > 10$.

115. Let $A = \{x: |x - \mu| \geq k\sigma\}$. Then $\sigma^2 = \sum_A (x - \mu)^2 p(x) \geq (k\sigma)^2 \sum_A p(x)$. But

$\sum_A p(x) = P(X \text{ is in } A) = P(|X - \mu| \geq k\sigma)$, so $\sigma^2 \geq k^2\sigma^2 \cdot P(|X - \mu| \geq k\sigma)$, as desired.

Chapter 3: Discrete Random Variables and Probability Distributions

CHAPTER 4

Section 4.1

1.

   a. $P(x \le 1) = \int_{-\infty}^{1} f(x)dx = \int_{0}^{1} \frac{1}{2}x\,dx = \frac{1}{4}x^2 \Big]_{0}^{1} = .25$

   b. $P(.5 \le X \le 1.5) = \int_{.5}^{1.5} \frac{1}{2}x\,dx = \frac{1}{4}x^2 \Big]_{.5}^{1.5} = .5$

   c. $P(x > 1.5) = \int_{1.5}^{\infty} f(x)dx = \int_{1.5}^{2} \frac{1}{2}x\,dx = \frac{1}{4}x^2 \Big]_{1.5}^{2} = \frac{7}{16} \approx .438$

3.

   a.   Graph of $f(x) = .09375(4 - x^2)$

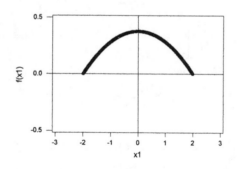

   b. $P(X > 0) = \int_{0}^{2} .09375(4 - x^2)dx = .09375(4x - \frac{x^3}{3}) \Big]_{0}^{2} = .5$

   c. $P(-1 < X < 1) = \int_{-1}^{1} .09375(4 - x^2)dx = .6875$

   d. $P(x < -.5 \text{ OR } x > .5) = 1 - P(-.5 \le X \le .5) = 1 - \int_{-.5}^{.5} .09375(4 - x^2)dx$

            $= 1 - .3672 = .6328$

81

5.

a. $1 = \int_{-\infty}^{\infty} f(x)dx = \int_{0}^{2} kx^2 dx = k\left(\frac{x^3}{3}\right)\Big|_{0}^{2} = k\left(\frac{8}{3}\right) \Rightarrow k = \frac{3}{8}$

b. $P(0 \le X \le 1) = \int_{0}^{1} \frac{3}{8}x^2 dx = \frac{1}{8}x^3\Big|_{0}^{1} = \frac{1}{8} = .125$

c. $P(1 \le X \le 1.5) = \int_{1}^{1.5} \frac{3}{8}x^2 dx = \frac{1}{8}x^3\Big|_{1}^{1.5} = \frac{1}{8}\left(\frac{3}{2}\right)^3 - \frac{1}{8}(1)^3 = \frac{19}{64} \approx .2969$

d. $P(X \ge 1.5) = 1 - \int_{0}^{1.5} \frac{3}{8}x^2 dx = \frac{1}{8}x^3\Big|_{0}^{1.5} = 1 - \left[\frac{1}{8}\left(\frac{3}{2}\right)^3 - 0\right] = 1 - \frac{27}{64} = \frac{37}{64} \approx .5781$

7.

a. $f(x) = \frac{1}{10}$ for $25 \le x \le 35$ and $= 0$ otherwise

b. $P(X > 33) = \int_{33}^{35} \frac{1}{10} dx = .2$

c. $E(X) = \int_{25}^{35} x \cdot \frac{1}{10} dx = \frac{x^2}{20}\Big|_{25}^{35} = 30$

30 ± 2 is from 28 to 32 minutes:

$P(28 < X < 32) = \int_{28}^{32} \frac{1}{10} dx = \frac{1}{10}x\Big|_{28}^{32} = .4$

d. $P(a \le x \le a+2) = \int_{a}^{a+2} \frac{1}{10} dx = .2$, since the interval has length 2.

9.

a. $P(X \le 6) = = \int_{5}^{6} .15e^{-.15(x-5)} dx = .15\int_{0}^{.5} e^{-.15u} du$ (after $u = x - .5$)

$= e^{-.15u}\Big|_{0}^{5.5} = 1 - e^{-.825} \approx .562$

b. $1 - .562 = .438; .438$

c. $P(5 \le Y \le 6) = P(Y \le 6) - P(Y \le 5) \approx .562 - .491 = .071$

Section 4.2

11.

    a. $P(X \le 1) = F(1) = \frac{1}{4} = .25$

    b. $P(.5 \le X \le 1) = F(1) - F(.5) = \frac{3}{16} = .1875$ $\frac{1}{2} = F(\tilde{\mu}) = \frac{1}{2}$

    c. $P(X > .5) = 1 - P(X \le .5) = 1 - F(.5) = \frac{15}{16} = .9375$

    d. $.5 = F(\tilde{\mu}) = \dfrac{\tilde{\mu}^2}{4} \Rightarrow \tilde{\mu}^2 = 2 \Rightarrow \tilde{\mu} = \sqrt{2} \approx 1.414$

    e. $f(x) = F'(x) = \frac{x}{2}$ for $0 \le x < 2$, and $= 0$ otherwise

    f. $E(X) = \displaystyle\int_{-\infty}^{\infty} x \cdot f(x)dx = \int_{0}^{2} x \cdot \frac{1}{2} x dx = \frac{1}{2}\int_{0}^{2} x^2 dx = \left.\frac{x^3}{6}\right]_{0}^{2} = \frac{8}{6} \approx 1.333$

    g. $E(X^2) = \displaystyle\int_{-\infty}^{\infty} x^2 f(x)dx = \int_{0}^{2} x^2 \frac{1}{2} x dx = \frac{1}{2}\int_{0}^{2} x^3 dx = \left.\frac{x^4}{8}\right]_{0}^{2} = 2,$

    So $Var(X) = E(X^2) - [E(X)]^2 = 2 - \left(\frac{8}{6}\right)^2 = \frac{8}{36} \approx .222$, $\sigma_x \approx .471$

    h. From **g**, $E(X^2) = 2$

13.

a. $1 = \int_1^\infty \dfrac{k}{x^4}dx \Rightarrow 1 = \dfrac{-k}{3}x^{-3}\Big|_1^\infty \Rightarrow 1 = 0 - (-\dfrac{k}{3})(1) \Rightarrow 1 = \dfrac{k}{3} \Rightarrow k = 3$

b. cdf: $F(x) = \int_{-\infty}^x f(y)dy = \int_1^x 3y^{-4}dy = -\dfrac{3}{3}y^{-3}\Big|_1^x = -x^{-3} + 1 = 1 - \dfrac{1}{x^3}$. So

$$F(x) = \begin{cases} 0, & x \le 1 \\ 1 - x^{-3}, & x > 1 \end{cases}$$

c. $P(x > 2) = 1 - F(2) = 1 - \left(1 - \frac{1}{8}\right) = \frac{1}{8}$ or .125;

$P(2 < x < 3) = F(3) - F(2) = \left(1 - \frac{1}{27}\right) - \left(1 - \frac{1}{8}\right) = .963 - .875 = .088$

d. $E(x) = \int_1^\infty x\left(\dfrac{3}{x^4}\right)dx = \int_1^\infty \left(\dfrac{3}{x^3}\right)dx = -\dfrac{3}{2}x^{-2}\Big|_1^x = 0 + \dfrac{3}{2} = \dfrac{3}{2}$

$E(x^2) = \int_1^\infty x^2\left(\dfrac{3}{x^4}\right)dx = \int_1^\infty \left(\dfrac{3}{x^2}\right)dx = -3x^{-1}\Big|_1^x = 0 + 3 = 3$

$V(x) = E(x^2) - [E(x)]^2 = 3 - \left(\dfrac{3}{2}\right)^2 = 3 - \dfrac{9}{4} = \dfrac{3}{4}$ or .75

$\sigma = \sqrt{V(x)} = \sqrt{\tfrac{3}{4}} = .866$

e. $P(1.5 - .866 < x < 1.5 + .866) = P(x < 2.366) = F(2.366) =$
$= 1 - (2.366^{-3}) = .9245$

15.

a. $F(X) = 0$ for $x \le 0$, $= 1$ for $x \ge 1$, and for $0 < X < 1$,

$$F(X) = \int_{-\infty}^{x} f(y)dy = \int_{0}^{x} 90y^{8}(1-y)dy = 90\int_{0}^{x}(y^{8} - y^{9})dy$$

$$90\left(\tfrac{1}{9}y^{9} - \tfrac{1}{10}y^{10}\right)\Big|_{0}^{x} = 10x^{9} - 9x^{10}$$

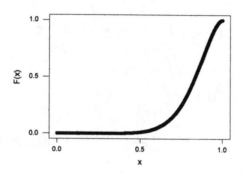

b. $F(.5) = 10(.5)^{9} - 9(.5)^{10} \approx .0107$

c. $P(.25 \le X \le .5) = F(.5) - F(.25) \approx .0107 - [10(.25)^{9} - 9(.25)^{10}]$
$$\approx .0107 - .0000 \approx .0107$$

d. The 75th percentile is the value of x for which $F(x) = .75$
$$\Rightarrow .75 = 10(x)^{9} - 9(x)^{10} \qquad \Rightarrow x \approx .9036$$

e. $E(X) = \int_{-\infty}^{\infty} x \cdot f(x)dx = \int_{0}^{1} x \cdot 90x^{8}(1-x)dx = 90\int_{0}^{1} x^{9}(1-x)dx$

$$= 9x^{10} - \tfrac{90}{11}x^{11}\Big|_{0}^{1} = \tfrac{9}{11} \approx .8182$$

$E(X^{2}) = \int_{-\infty}^{\infty} x^{2} \cdot f(x)dx = \int_{0}^{1} x^{2} \cdot 90x^{8}(1-x)dx = 90\int_{0}^{1} x^{10}(1-x)dx$

$$= \tfrac{90}{11}x^{11} - \tfrac{90}{12}x^{12}\Big|_{0}^{1} \approx .6818$$

$V(X) \approx .6818 - (.8182)^{2} = .0124, \qquad \sigma_{x} = .11134.$

f. $\mu \pm \sigma = (.7068, .9295)$. Thus, $P(\mu - \sigma \le X \le \mu + \sigma) = F(.9295) - F(.7068)$
$$= .8465 - .1602 = .6863$$

17.

a. For $2 \le X \le 4$, $F(X) = \int_{-\infty}^{x} f(y)dy = \int_{2}^{x} \frac{3}{4}[1-(y-3)^2]dy$ (let $u = $ y-3)

$$= \int_{-1}^{x-3} \frac{3}{4}[1-u^2]du = \frac{3}{4}\left[u - \frac{u^3}{3}\right]_{-1}^{x-3} = \frac{3}{4}\left[x - \frac{7}{3} - \frac{(x-3)^3}{3}\right]. \text{ Thus}$$

$$F(x) = \begin{cases} 0 & x < 2 \\ \frac{1}{4}[3x - 7 - (x-3)^3] & 2 \le x \le 4 \\ 1 & x > 4 \end{cases}$$

b. By symmetry of f(x), $\tilde{\mu} = 3$

c. E(X) = $\int_{2}^{4} x \cdot \frac{3}{4}[1-(x-3)^2]dx = \frac{3}{4}\int_{-1}^{1}(y+3)(1-y^2)dx$

$$= \frac{3}{4}\left[3y + \frac{y^2}{2} - y^3 - \frac{y^4}{4}\right]_{-1}^{1} = \frac{3}{4} \cdot 4 = 3$$

$$V(X) = \int_{-\infty}^{\infty}(x-\mu)^2 f(x)dx = \frac{3}{4}\int_{2}^{4}(x-3)^2 \cdot [1-(x-3)^2]dx$$

$$= \frac{3}{4}\int_{-1}^{1} y^2(1-y^2)dy = \frac{3}{4} \cdot \frac{4}{15} = \frac{1}{5} = .2$$

19.

a. $P(X \le 1) = F(1) = .25[1 + \ln(4)] \approx .597$

b. $P(1 \le X \le 3) = F(3) - F(1) \approx .966 - .597 \approx .369$

c. $f(x) = F'(x) = .25 \ln(4) - .25 \ln(x)$ for $0 < x < 4$

21. $E(area) = E(\pi R^2) = \int_{-\infty}^{\infty} \pi r^2 f(r)dr = \int_{9}^{11} \pi r^2 \left(\frac{3}{4}\right)\left(1-(10-r)^2\right)dr$

$= \left(\frac{3}{4}\right)\pi \int_{9}^{11} r^2 \left(1-(100-20r+r^2)\right)dr = \frac{3}{4}\pi \int_{9}^{11} -99r^2 + 20r^3 - r^4 dr = 100 \cdot 2\pi$

$E(X) = \int_{-\infty}^{\infty} x \cdot f(x)dx$

23.   With X = temperature in °C, temperature in °F $= \dfrac{9}{5}X + 32$, so

$E\left[\frac{9}{5}X + 32\right] = \frac{9}{5}(120) + 32 = 248, \quad Var\left[\frac{9}{5}X + 32\right] = \left(\frac{9}{5}\right)^2 \cdot (2)^2 = 12.96,$

so σ = 3.6

25.

   a. $P(Y \le 1.8\tilde{\mu} + 32) = P(1.8X + 32 \le 1.8\tilde{\mu} + 32) = P(X \le \tilde{\mu}) = .5$

   b. 90th for Y = 1.8η(.9) + 32 where η(.9) is the 90th percentile for X, since
   $P(Y \le 1.8η(.9) + 32) = P(1.8X + 32 \le 1.8η(.9) + 32)$
   $= (X \le η(.9)) = .9$ as desired.

   c.   The (100p)th percentile for Y is 1.8η(p) + 32, verified by substituting p for .9 in the argument of **b**. When Y = aX + b, (i.e. a linear transformation of X), and the (100p)th percentile of the X distribution is η(p), then the corresponding (100p)th percentile of the Y distribution is a·η(p) + b. (same linear transformation applied to X's percentile)

Section 4.3

27.

    a. .9838 is found in the 2.1 row and the .04 column of the standard normal table so c = 2.14.

    b. $P(0 \leq Z \leq c) = .291 \Rightarrow \Phi(c) = .7910 \Rightarrow c = .81$

    c. $P(c \leq Z) = .121 \Rightarrow 1 - P(c \leq Z) = P(Z < c) = \Phi(c) = 1 - .121 = .8790 \Rightarrow c = 1.17$

    d. $P(-c \leq Z \leq c) = \Phi(c) - \Phi(-c) = \Phi(c) - (1 - \Phi(c)) = 2\Phi(c) - 1$
       $\Rightarrow \Phi(c) = .9920 \Rightarrow c = .97$

    e. $P(c \leq |Z|) = .016 \Rightarrow 1 - .016 = .9840 = 1 - P(c \leq |Z|) = P(|Z| < c)$
       $= P(-c < Z < c) = \Phi(c) - \Phi(-c) = 2\Phi(c) - 1$
       $\Rightarrow \Phi(c) = .9920 \Rightarrow c = 2.41$

29.

    a. Area under Z curve above $z_{.0055}$ is .0055, which implies that
       $\Phi(z_{.0055}) = 1 - .0055 = .9945$, so $z_{.0055} = 2.54$

    b. $\Phi(z_{.09}) = .9100 \Rightarrow z = 1.34$ (since .9099 appears as the 1.34 entry).

    c. $\Phi(z_{.633}) = $ area below $z_{.633} = .3370 \Rightarrow z_{.633} \approx -.42$

31.

    a. $P(X \leq 18) = P\left(z \leq \dfrac{18 - 15}{1.25}\right) = P(Z \leq 2.4) = \Phi(2.4) = .9452$

    b. $P(10 \leq X \leq 12) = P(-4.00 \leq Z \leq -2.40) \approx P(Z \leq -2.40) = \Phi(-2.40) = .0082$

    c. $P(|X - 10| \leq 2(1.25)) = P(-2.50 \leq X-15 \leq 2.50) = P(12.5 \leq X \leq 17.5)$
                            $P(-2.00 \leq Z \leq 2.00) = .9544$

Chapter 4: Continuous Random Variables and Probability Distributions

33.

    a. $P(X \geq 10) = P(Z \geq .43) = 1 - \Phi(.43) = 1 - .6664 = .3336$.

        $P(X > 10) = P(X \geq 10) = .3336$, since for any continuous distribution, $P(x = a) = 0$.

    b. $P(X > 20) = P(Z > 4) \approx 0$

    c. $P(5 \leq X \leq 10) = P(-1.36 \leq Z \leq .43) = \Phi(.43) - \Phi(-1.36) = .6664 - .0869 = .5795$

    d. $P(8.8 - c \leq X \leq 8.8 + c) = .98$, so $8.8 - c$ and $8.8 + c$ are at the 1st and the 99th percentile of the given distribution, respectively. The 1st percentile of the standard normal distribution has the value -2.33, so

        $8.8 - c = \mu + (-2.33)\sigma = 8.8 - 2.33(2.8) \Rightarrow c = 2.33(2.8) = 6.524$.

    e. From a, $P(x > 10) = .3336$. Define event A as {diameter > 10}, then P(at least one A_i) $= 1 - P(\text{no } A_i) = 1 - P(A')^4 = 1 - (1 - .3336)^4 = 1 - .1972 = .8028$

35.

    a. $\mu + \sigma \cdot (91^{st}$ percentile from std normal$) = 30 + 5(1.34) = 36.7$

    b. $30 + 5(-1.555) = 22.225$

    c. $\mu = 3.000 \ \mu m$; $\sigma = 0.140$. We desire the 90th percentile: $30 + 1.28(0.14) = 3.179$

37. $P(\text{damage}) = P(X < 100) = P\left(z < \dfrac{100 - 200}{300}\right) = P(Z < -3.33) = .0004$

    $P(\text{at least one among five is damaged})$ $= 1 - P(\text{none damaged})$

                                    $= 1 - (.9996)^5 = 1 - .998 = .002$

39. Since 1.28 is the 90th z percentile ($z_{.1} = 1.28$) and -1.645 is the 5th z percentile ($z_{.05} = 1.645$), the given information implies that $\mu + \sigma(1.28) = 10.256$ and $\mu + \sigma(-1.645) = 9.671$, from which $\sigma(-2.925) = -.585$, $\sigma = .2000$, and $\mu = 10$.

41.   With $\mu = .500$ inches, the acceptable range for the diameter is between .496 and .504 inches, so unacceptable bearings will have diameters smaller than .496 or larger than .504. The new distribution has $\mu = .499$ and $\sigma = .002$. P(x < .496 or x >.504) =

$$P\left(z < \frac{.496 - .499}{.002}\right) + P\left(z > \frac{.504 - .499}{.002}\right) = P(z < -1.5) + P(z > 2.5)$$

$\Phi(-1.5) + (1 - \Phi(2.5)) = .0068 + .0062 = .073$, or 7.3% of the bearings will be unacceptable.

43.   The stated condition implies that 99% of the area under the normal curve with $\mu = 10$ and $\sigma = 2$ is to the left of $c - 1$, so $c - 1$ is the 99th percentile of the distribution. Thus $c - 1 = \mu + \sigma(2.33) = 20.155$, and $c = 21.155$.

45.   X ~N(3432, 482)

a. $P(x > 4000) = P\left(Z > \dfrac{4000 - 3432}{482}\right) = P(z > 1.18)$

$= 1 - \Phi(1.18) = 1 - .8810 = .1190$

$P(3000 < x < 4000) = P\left(\dfrac{3000 - 3432}{482} < Z < \dfrac{4000 - 3432}{482}\right)$

$= \Phi(1.18) - \Phi(-.90) = .8810 - .1841 = .6969$

b. $P(x < 2000 \, or \, x > 5000) = P\left(Z < \dfrac{2000 - 3432}{482}\right) + P\left(Z > \dfrac{5000 - 3432}{482}\right)$

$= \Phi(-2.97) + [1 - \Phi(3.25)] = .0015 + .0006 = .0021$

c.   We will use the conversion 1 lb = 454 g, then 7 lbs = 3178 grams, and we wish to find $P(x > 3178) = P\left(Z > \dfrac{3178 - 3432}{482}\right) = 1 - \Phi(-.53) = .7019$

d. We need the top .0005 and the bottom .0005 of the distribution. Using the Z table, both .9995 and .0005 have multiple z values, so we will use a middle value, ±3.295. Then 3432±(482)3.295 = 1844 and 5020, or the most extreme .1% of all birth weights are less than 1844 g and more than 5020 g.

e. Converting to lbs yields mean 7.5595 and s.d. 1.0608. Then

$$P(x > 7) = P\left(Z > \frac{7-7.5595}{1.0608}\right) = 1 - \Phi(-.53) = .7019$$ This yields the same

answer as in part c.

47. $P(\,|X - \mu\,|\geq \sigma\,) = P(\,X \leq \mu - \sigma \text{ or } X \geq \mu + \sigma\,)$
            $= 1 - P(\mu - \sigma \leq X \leq \mu + \sigma) = 1 - P(-1 \leq Z \leq 1) = .3174$
    Similarly, $P(\,|X - \mu\,|\geq 2\sigma\,) = 1 - P(-2 \leq Z \leq 2) = .0456$
            And $P(\,|X - \mu\,|\geq 3\sigma\,) = 1 - P(-3 \leq Z \leq 3) = .0026$

49.

P:	.5	.6	.8
μ:	12.5	15	20
σ:	2.50	2.45	2.00

a.

P(15≤ X ≤20)		P(14.5 ≤ normal ≤ 20.5)
.5	.212	P(.80 ≤ Z ≤ 3.20) = .2112
.6	.577	P(-.20 ≤ Z ≤ 2.24) = .5668
.8	.573	P(-2.75 ≤ Z ≤ .25) = .5957

b.

P(X ≤15)	P(normal ≤ 15.5)
.885	P(Z ≤ 1.20) = .8849
.575	P(Z ≤ .20) = .5793
.017	P(Z ≤ -2.25) = .0122

c.

P(20 ≤X)	P(19.5 ≤ normal)
.002	.0026
.029	.0329
.617	.5987

51.   N = 500, p = .4, μ = 200, σ = 10.9545

    a. P(180 ≤ X ≤ 230) = P(179.5 ≤ normal ≤ 230.5) = P(-1.87 ≤ Z ≤ 2.78) = .9666

    b. P(X < 175) = P(X ≤ 174) = P(normal ≤ 174.5) = P(Z ≤ -2.33) = .0099

53.

    a. $F_y(y) = P(Y \le y) = P(aX + b \le y) = P\left(X \le \dfrac{(y-b)}{a}\right)$ (for a > 0).

    Now differentiate with respect to y to obtain

$$f_y(y) = F_y'(y) = \frac{1}{\sqrt{2\pi}a\sigma}e^{-\frac{1}{2a^2\sigma^2}[y-(a\mu+b)]^2}$$ so Y is normal with mean aμ + b

    and variance $a^2\sigma^2$.

    b. Normal, mean $\frac{9}{5}(115) + 32 = 239$, variance = 12.96

Section 4.4

55.

    a. $\Gamma(6) = 5! = 120$

    b. $\Gamma\left(\dfrac{5}{2}\right) = \dfrac{3}{2}\Gamma\left(\dfrac{1}{2}\right) = \dfrac{3}{2}\cdot\dfrac{1}{2}\cdot\Gamma\left(\dfrac{1}{2}\right) = \left(\dfrac{3}{4}\right)\sqrt{\pi} \approx 1.329$

    c. $F(4;5) = .371$ from row 4, column 5 of Table A.4

    d. $F(5;4) = .735$

    e. $F(0;4) = P(X \le 0; \alpha = 4) = 0$

57.

    a. $\mu = 20, \ \sigma^2 = 80 \Rightarrow \alpha\beta = 20, \ \alpha\beta^2 = 80 \Rightarrow \beta = \frac{80}{20}, \ \alpha = 5$

    b. $P(X \le 24) = F\left(\frac{24}{4};5\right) = F(6;5) = .715$

    c. $P(20 \le X \le 40) = F(10;5) - F(5;5) = .411$

59.

    a. $E(X) = \dfrac{1}{\lambda} = 1$

    b. $\sigma = \dfrac{1}{\lambda} = 1$

    c. $P(X \le 4) = 1 - e^{-(1)(4)} = 1 - e^{-4} = .982$

    d. $P(2 \le X \le 5) = 1 - e^{-(1)(5)} - \left[1 - e^{-(1)(2)}\right] = e^{-2} - e^{-5} = .129$

61.   Mean $= \dfrac{1}{\lambda} = 25{,}000$ implies $\lambda = .00004$

a. $P(X > 20{,}000) = 1 - P(X \le 20{,}000) = 1 - F(20{,}000; .00004)$
$= e^{-(.00004)(20{,}000)} = .449$

$P(X \le 30{,}000) = F(30{,}000; .00004) = e^{-1.2} = .699$
$P(20{,}000 \le X \le 30{,}000) = .699 - .551 = .148$

b. $\sigma = \dfrac{1}{\lambda} = 25{,}000$, so $P(X > \mu + 2\sigma) = P(\,x > 75{,}000) =$

$1 - F(75{,}000; .00004) = .05.$
Similarly, $P(X > \mu + 3\sigma) = P(\,x > 100{,}000) = .018$

63.

a. $\{X \ge t\} = A_1 \cap A_2 \cap A_3 \cap A_4 \cap A_5$

b. $P(X \ge t) = P(\,A_1\,) \cdot P(\,A_2\,) \cdot P(\,A_3\,) \cdot P(\,A_4\,) \cdot P(\,A_5\,) = \left(e^{-\lambda t}\right)^5 = e^{-.05t}$, so $F_x(t) =$
$P(X \le t) = 1 - e^{-.05t}$, $f_x(t) = .05 e^{-.05t}$ for $t \ge 0$. Thus X also ha an exponential
distribution , but with parameter $\lambda = .05.$

c.   By the same reasoning, $P(X \le t) = 1 - e^{-n\lambda t}$, so X has an exponential distribution
with parameter $n\lambda$.

65.

a. $\{X^2 \le y\} = \left\{-\sqrt{y} \le X \le \sqrt{y}\right\}$

b. $P(X^2 \le y) = \displaystyle\int_{-\sqrt{y}}^{\sqrt{y}} \dfrac{1}{\sqrt{2\pi}} e^{-z^2/2}\,dz$. Now differentiate with respect to y to obtain the

chi-squared p.d.f. with $\nu = 1.$

Section 4.5

67.

    a. $P(X \le 250) = F(250;2.5, 200) = 1 - e^{-(250/200)^{2.5}} = 1 - e^{-1.75} \approx .8257$

        $P(X < 250) = P(X \le 250) \approx .8257$

        $P(X > 300) = 1 - F(300; 2.5, 200) = e^{-(1.5)^{2.5}} = .0636$

    b. $P(100 \le X \le 250) = F(250;2.5, 200) - F(100;2.5, 200) \approx .8257 - .162 = .6637$

    c. The median $\tilde{\mu}$ is requested. The equation $F(\tilde{\mu}) = .5$ reduces to

        $.5 = e^{-(\tilde{\mu}/200)^{2.5}}$, i.e., $\ln(.5) \approx -\left(\dfrac{\tilde{\mu}}{200}\right)^{2.5}$, so $\tilde{\mu} = (.6931)^{.4}(200) = 172.727.$

69. $\mu = \displaystyle\int_0^\infty x \cdot \dfrac{\alpha}{\beta^\alpha} x^{\alpha-1} e^{-(x/\beta)^\alpha} \, dx$ = (after $y = \left(\dfrac{x}{\beta}\right)^\alpha$, $dy = \dfrac{\alpha x^{\alpha-1}}{\beta^\alpha} dx$)

        $\beta \displaystyle\int_0^\infty y^{1/\alpha} e^{-y} \, dy = \beta \cdot \Gamma\left(1 + \dfrac{1}{\alpha}\right)$ by definition of the gamma function.

71.   X ~ Weibull: $\alpha = 20, \beta = 100$

    a. $F(x, 20, \beta) = 1 - e^{-\left(\frac{x}{\beta}\right)^\alpha} = 1 - e^{-\left(\frac{105}{100}\right)^{20}} = 1 - .070 = .930$

    b. $F(105) - F(100) = .930 - (1 - e^{-1}) = .930 - .632 = .298$

    c. $.50 = 1 - e^{-\left(\frac{x}{100}\right)^{20}} \Rightarrow e^{-\left(\frac{x}{100}\right)^{20}} = .50 \Rightarrow -\left(\frac{x}{100}\right)^{20} = \ln(.50)$

        $\left(\dfrac{-x}{100}\right) = \sqrt[20]{\ln(.50)} \Rightarrow -x = 100\left(\sqrt[20]{\ln(.50)}\right) \Rightarrow x = 98.18$

73.

   a. $E(X) = e^{3.5+(1.2)^2/2} = 68.0335$; $V(X) = e^{2(3.5)+(1.2)^2} \cdot \left(e^{(1.2)^2} - 1\right) = 14907.168$;

       $\sigma_x = 122.0949$

   b. $P(50 \le X \le 250) = P\left(z \le \dfrac{\ln(250)-3.5}{1.2}\right) - P\left(z \le \dfrac{\ln(50)-3.5}{1.2}\right)$

       $P(Z \le 1.68) - P(Z \le .34) = .9535 - .6331 = .3204.$

   c. $P(X \le 68.0335) = P\left(z \le \dfrac{\ln(68.0335)-3.5}{1.2}\right) = P(Z \le .60) = .7257.$ The

       lognormal distribution is not a symmetric distribution.

75.

   a. $E(X) = e^{5+(.01)/2} = e^{5.005} = 149.157$; $Var(X) = e^{10+(.01)} \cdot \left(e^{.01} - 1\right) = 223.594$

   b. $P(X > 125) = 1 - P(X \le 125) =$

       $= 1 - P\left(z \le \dfrac{\ln(125)-5}{.1}\right) = 1 - \Phi(-1.72) = .9573$

   c. $P(110 \le X \le 125) = \Phi(-1.72) - \Phi\left(\dfrac{\ln(110)-5}{.1}\right) = .0427 - .0013 = .0414$

   d. $\tilde{\mu} = e^5 = 148.41$ (continued)

   e. P(any particular one has $X > 125$) = .9573 $\Rightarrow$ expected # = 10(.9573) = 9.573

   f. We wish the 5th percentile, which is $e^{5+(-1.645)(.1)} = 125.90$

77.   The point of symmetry must be $\frac{1}{2}$, so we require that $f\left(\frac{1}{2} - \mu\right) = f\left(\frac{1}{2} + \mu\right)$, i.e.,

     $\left(\frac{1}{2} - \mu\right)^{\alpha-1}\left(\frac{1}{2} + \mu\right)^{\beta-1} = \left(\frac{1}{2} + \mu\right)^{\alpha-1}\left(\frac{1}{2} - \mu\right)^{\beta-1}$, which in turn implies that $\alpha = \beta$.

79.

a. $E(X) = \int_0^1 x \cdot \frac{\Gamma(\alpha + \beta)}{\Gamma(\alpha)\Gamma(\beta)} x^{\alpha-1}(1-x)^{\beta-1} dx = \frac{\Gamma(\alpha + \beta)}{\Gamma(\alpha)\Gamma(\beta)} \int_0^1 x^{\alpha}(1-x)^{\beta-1} dx$

$\frac{\Gamma(\alpha + \beta)}{\Gamma(\alpha)\Gamma(\beta)} \cdot \frac{\Gamma(\alpha +1)\Gamma(\beta)}{\Gamma(\alpha + \beta +1)} = \frac{\alpha\Gamma(\alpha)}{\Gamma(\alpha)\Gamma(\beta)} \cdot \frac{\Gamma(\alpha + \beta)}{(\alpha + \beta)\Gamma(\alpha + \beta)} = \frac{\alpha}{\alpha + \beta}$

b. $E[(1-X)^m] = \int_0^1 (1-x)^m \cdot \frac{\Gamma(\alpha + \beta)}{\Gamma(\alpha)\Gamma(\beta)} x^{\alpha-1}(1-x)^{\beta-1} dx$

$= \frac{\Gamma(\alpha + \beta)}{\Gamma(\alpha)\Gamma(\beta)} \int_0^1 x^{\alpha-1}(1-x)^{m+\beta-1} dx = \frac{\Gamma(\alpha + \beta) \cdot \Gamma(m + \beta)}{\Gamma(\alpha + \beta + m)\Gamma(\beta)}$

For m = 1, $E(1-X) = \frac{\beta}{\alpha + \beta}$.

Section 4.6

81. The given probability plot is quite linear, and thus it is quite plausible that the tension distribution is normal.

83. The z percentile values are as follows: -1.86, -1.32, -1.01, -0.78, -0.58, -0.40, -0.24,-0.08, 0.08, 0.24, 0.40, 0.58, 0.78, 1.01, 1.30, and 1.86. The accompanying probability plot is reasonably straight, and thus it would be reasonable to use estimating methods that assume a normal population distribution.

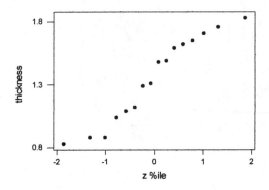

85. The (z percentile, observation) pairs are (-1.66, .736), (-1.32, .863),   (-1.01, .865), (-.78, .913), (-.58, .915), (-.40, .937), (-.24, .983), (-.08, 1.007), (.08, 1.011), (.24, 1.064), (.40, 1.109), (.58, 1.132), (.78, 1.140), (1.01, 1.153), (1.32, 1.253), (1.86, 1.394). The accompanying probability plot is very straight, suggesting that an assumption of population normality is extremely plausible.

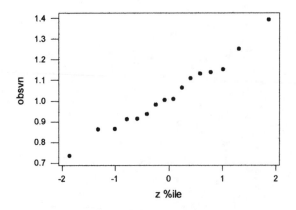

87.   To check for plausibility of a lognormal population distribution for the rainfall data of Exercise 81 in Chapter 1, take the natural logs and construct a normal probability plot. This plot and a normal probability plot for the original data appear below. Clearly the log transformation gives quite a straight plot, so lognormality is plausible. The curvature in the plot for the original data implies a positively skewed population distribution - like the lognormal distribution.

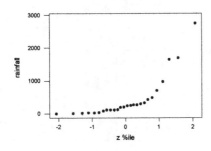 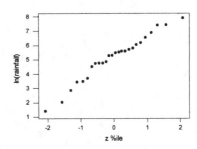

89.   The pattern in the plot (below, generated by Minitab) is quite linear. It is very plausible that strength is normally distributed.

Normal Probability Plot

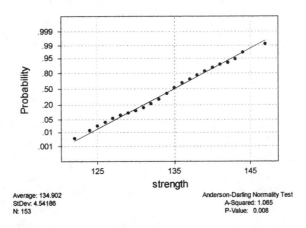

Average: 134.902
StDev: 4.54186
N: 153

Anderson-Darling Normality Test
A-Squared: 1.065
P-Value: 0.008

91. The $(100p)^{th}$ percentile $\eta(p)$ for the exponential distribution with $\lambda = 1$ satisfies $F(\eta(p)) = 1 - \exp[-\eta(p)] = p$, i.e., $\eta(p) = -\ln(1-p)$. With $n = 16$, we need $\eta(p)$ for $p = \frac{.5}{16}, \frac{1.5}{16}, ..., \frac{15.5}{16}$. These are .032, .398, .170, .247, .330, .421, .521, .633, .758, .901, 1.068, 1.269, 1.520, 1.856, 2.367, 3.466. this plot exhibits substantial curvature, casting doubt on the assumption of an exponential population distribution. Because λ is a scale parameter (as is σ for the normal family), $\lambda = 1$ can be used to assess the plausibility of the entire exponential family.

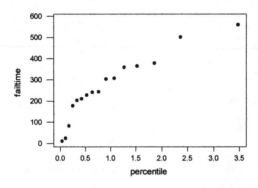

Supplementary Exercises

93.

a. For $0 \le Y \le 25$, $F(y) = \dfrac{1}{24} \int_0^y \left(u - \dfrac{u^2}{12} \right) = \dfrac{1}{24} \left(\dfrac{u^2}{2} - \dfrac{u^3}{36} \right) \Big]_0^y$. Thus

$$F(y) = \begin{cases} 0 & y < 0 \\ \dfrac{1}{48}\left(y^2 - \dfrac{y^3}{18} \right) & 0 \le y \le 12 \\ 1 & y > 12 \end{cases}$$

b. $P(Y \le 4) = F(4) = .259$, $P(Y > 6) = 1 - F(6) = .5$
$P(4 \le X \le 6) = F(6) - F(4) = .5 - .259 = .241$

c. $E(Y) = \dfrac{1}{24} \int_0^{12} y^2 \left(1 - \dfrac{y}{12} \right) dy = \dfrac{1}{24} \left[\dfrac{y^3}{3} - \dfrac{y^4}{48} \right]_0^{12} = 6$

$E(Y^2) = \dfrac{1}{24} \int_0^{12} y^3 \left(1 - \dfrac{y}{12} \right) dy = 43.2$, so $V(Y) = 43.2 - 36 = 7.2$

d. $P(Y < 4 \text{ or } Y > 8) = 1 - P(4 \le X \le 8) = .518$

e. the shorter segment has length $\min(Y, 12 - Y)$ so
$E[\min(Y, 12 - Y)] = \int_0^{12} \min(y, 12 - y) \cdot f(y) dy = \int_0^6 \min(y, 12 - y) \cdot f(y) dy$
$+ \int_6^{12} \min(y, 12 - y) \cdot f(y) dy = \int_0^6 y \cdot f(y) dy + \int_6^{12} (12 - y) \cdot f(y) dy =$
$\dfrac{90}{24} = .3.75$

95.

a. By differentiation,

$$f(x) = \begin{cases} x^2 & 0 \le x < 1 \\ \dfrac{7}{4} - \dfrac{3}{4}x & 1 \le y \le \dfrac{7}{3} \\ 0 & otherwise \end{cases}$$

b. $P(.5 \le X \le 2) = F(2) - F(.5) = 1 - \dfrac{1}{2}\left(\dfrac{7}{3} - 2\right)\left(\dfrac{7}{4} - \dfrac{3}{4} \cdot 2\right) - \dfrac{(.5)^3}{3} = \dfrac{11}{12} = .917$

c. $E(X) = \int_0^1 x \cdot x^2 \, dx + \int_1^{7/3} x \cdot \left(\dfrac{7}{4} - \dfrac{3}{4}x\right) dx = \dfrac{131}{108} = 1.213$

97. $\mu = 137.2$ oz.; $\sigma = 1.6$ oz

a. $P(X > 135) = 1 - \Phi\left(\dfrac{135 - 137.2}{1.6}\right) = 1 - \Phi(-1.38) = 1 - .0838 = .9162$

b. With Y = the number among ten that contain more than 135 oz,
Y ~ Bin(10, .9162, so $P(Y \ge 8) = b(8; 10, .9162) + b(9; 10, .9162)$
$+ b(10; 10, .9162) = .9549$.

c. $\mu = 137.2;\ \dfrac{135 - 137.2}{\sigma} = -1.65 \Rightarrow \sigma = 1.33$

99.

a. $P(X > 100) = 1 - \Phi\left(\dfrac{100 - 96}{14}\right) = 1 - \Phi(.29) = 1 - .6141 = .3859$

b. $P(50 < X < 80) = \Phi\left(\dfrac{80 - 96}{14}\right) - \Phi\left(\dfrac{50 - 96}{14}\right)$
$= \Phi(-1.5) - \Phi(-3.29) = .1271 - .0005 = .1266$.

c. a = 5th percentile = 96 + (-1.645)(14) = 72.97.
b = 95th percentile = 96 + (1.645)(14) = 119.03. The interval (72.97, 119.03)
contains the central 90% of all grain sizes.

101.

   a.

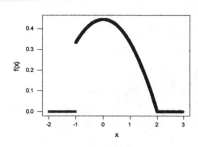

   b. $F(x) = 0$ for $x < -1$ or $== 1$ for $x > 2$. For $-1 \le x \le 2$,

$$F(x) = \int_{-1}^{x} \frac{1}{9}(4 - y^2)\,dy = \frac{1}{9}\left(4x - \frac{x^3}{3}\right) + \frac{11}{27}$$

   c. The median is 0 iff $F(0) = .5$. Since $F(0) = \frac{11}{27}$, this is not the case. Because $\frac{11}{27} < .5$, the median must be greater than 0.

   d. Y is a binomial r.v. with n = 10 and $p = P(X > 1) = 1 - F(1) = \frac{5}{27}$

103.

   a. $P(X \le 150) = \exp\left[-\exp\left(\dfrac{-(150 - 150)}{90}\right)\right] = \exp[-\exp(0)] = \exp(-1) = .368$,

      where $\exp(u) = e^u$. $P(X \le 300) = \exp[-\exp(-1.6667)] = .828$,

      and $P(150 \le X \le 300) = .828 - .368 = .460$.

   b. The desired value c is the 90th percentile, so c satisfies

      $.9 = \exp\left[-\exp\left(\dfrac{-(c - 150)}{90}\right)\right]$. Taking the natural log of each side twice in

      succession yields $\ln[\,\ln(.9)] = \dfrac{-(c - 150)}{90}$, so c = 90(2.250367) + 150 = 352.53.

   c. $f(x) = F'(X) = \dfrac{1}{\beta} \cdot \exp\left[-\exp\left(\dfrac{-(x - \alpha)}{\beta}\right)\right] \cdot \exp\left(\dfrac{-(x - \alpha)}{\beta}\right)$

d. We wish the value of x for which f(x) is a maximum; this is the same as the value of x for which ln[f(x)] is a maximum. The equation of $\dfrac{d[\ln(f(x))]}{dx} = 0$ gives

$$\exp\left(\dfrac{-(x-\alpha)}{\beta}\right) = 1, \text{ so } \dfrac{-(x-\alpha)}{\beta} = 0,$$ which implies that x = α. Thus the mode is α.

e. E(X) = .5772β + α = 201.95, whereas the mode is 150 and the median is −(90)ln[-ln(.5)] + 150 = 182.99. The distribution is positively skewed.

105.

a. From a graph of f(x; μ, σ) or by differentiation, x* = μ.

b. No; the density function has constant height for A ≤ X ≤ B.

c. F(x;λ) is largest for x = 0 (the derivative at 0 does not exist since f is not continuous there) so x* = 0.

d. $\ln[f(x;\alpha,\beta)] = -\ln(\beta^\alpha) - \ln(\Gamma(\alpha)) + (\alpha-1)\ln(x) - \dfrac{x}{\beta};$

$$\dfrac{d}{dx}\ln[f(x;\alpha,\beta)] = \dfrac{\alpha-1}{x} - \dfrac{1}{\beta} \Rightarrow x = x^* = (\alpha-1)\beta$$

e. From d $x^* = \left(\dfrac{\nu}{2}-1\right)(2) = \nu - 2.$

107.

a. Clearly $f(x; \lambda_1, \lambda_2, p) \geq 0$ for all x, and $\int_{-\infty}^{\infty} f(x; \lambda_1, \lambda_2, p)dx$

$$= \int_0^\infty \left[p\lambda_1 e^{-\lambda_1 x} + (1-p)\lambda_2 e^{-\lambda_2 x} \right] dx = p \int_0^\infty \lambda_1 e^{-\lambda_1 x} dx + (1-p) \int_0^\infty \lambda_2 e^{-\lambda_2 x} dx$$
$$= p + (1-p) = 1$$

b. For $x > 0$, $F(x; \lambda_1, \lambda_2, p)$

$$= \int_0^x f(y; \lambda_1, \lambda_2, p)dy = p(1 - e^{-\lambda_1 x}) + (1-p)(1 - e^{-\lambda_2 x}).$$

c. $E(X) = \int_0^\infty x \cdot \left[p\lambda_1 e^{-\lambda_1 x} + (1-p)\lambda_2 e^{-\lambda_2 x} \right] dx$

$$= p \int_0^\infty x\lambda_1 e^{-\lambda_1 x} dx + (1-p) \int_0^\infty x\lambda_2 e^{-\lambda_2 x} dx = \frac{p}{\lambda_1} + \frac{(1-p)}{\lambda_2}$$

d. $E(X^2) = \dfrac{2p}{\lambda_1^2} + \dfrac{2(1-p)}{\lambda_2^2}$, so $\text{Var}(X) = \dfrac{2p}{\lambda_1^2} + \dfrac{2(1-p)}{\lambda_2^2} - \left[\dfrac{p}{\lambda_1} + \dfrac{(1-p)}{\lambda_2} \right]^2$

e. For an exponential r.v., CV $= \dfrac{1/\lambda}{1/\lambda} = 1$. For X hyperexponential,

$$CV = \left[\frac{\frac{2p}{\lambda_1^2} + \frac{2(1-p)}{\lambda_2^2}}{\left[\frac{p}{\lambda_1} + \frac{(1-p)}{\lambda_2} \right]^2} - 1 \right]^{1/2} = \left[\frac{2(p\lambda_2^2 + (1-p)\lambda_1^2)}{(p\lambda_2 + (1-p)\lambda_1)^2} - 1 \right]^{1/2}$$

$= [2r - 1]^{1/2}$ where $r = \dfrac{(p\lambda_2^2 + (1-p)\lambda_1^2)}{(p\lambda_2 + (1-p)\lambda_1)^2}$. But straightforward algebra shows

that $r > 1$ provided $\lambda_1 \neq \lambda_2$, so that CV > 1.

f. $\mu = \dfrac{n}{\lambda}$, $\qquad \sigma^2 = \dfrac{n}{\lambda^2}$, $\qquad$ so $\sigma = \dfrac{\sqrt{n}}{\lambda}$ $\qquad$ and CV $= \dfrac{1}{\sqrt{n}} < 1$ if n > 1.

109.

a. $1 = \int_5^\infty \dfrac{k}{x^\alpha}\,dx = k \cdot \dfrac{5^{1-\alpha}}{\alpha - 1} \Rightarrow k = (\alpha - 1)5^{1-\alpha}$ where we must have $\alpha > 1$.

b. For $x \geq 5$, $F(x) = \int_5^x \dfrac{k}{y^\alpha}\,dy = 5^{1-\alpha}\left[\dfrac{1}{5^{1-\alpha}} - \dfrac{1}{x^{\alpha-1}}\right] = 1 - \left(\dfrac{5}{x}\right)^{\alpha-1}$.

c. $E(X) = \int_5^\infty x \cdot \dfrac{k}{x^\alpha}\,dx = \int_5^\infty x \cdot \dfrac{k}{x^{\alpha-1}}\,dx = \dfrac{k}{5^{\alpha-2}\cdot(\alpha-2)}$, provided $\alpha > 2$.

d. $P\left(\ln\left(\dfrac{X}{5}\right) \leq y\right) = P\left(\dfrac{X}{5} \leq e^y\right) = P(X \leq 5e^y) = F(5e^y) = 1 - \left(\dfrac{5}{5e^y}\right)^{\alpha-1}$

$1 - e^{-(\alpha-1)y}$, the cdf of an exponential r.v. with parameter α - 1.

111.

a.

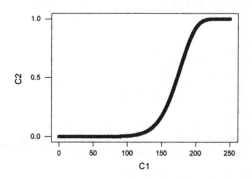

b. $P(X > 175) = 1 - F(175; 9, 180) = e^{-\left(\frac{175}{180}\right)^9} = .4602$

$P(150 \leq X \leq 175) = F(175; 9, 180) - F(150; 9, 180)$

$= .5398 - .1762 = .3636$

c. $P(\text{at least one}) = 1 - P(\text{none}) = 1 - (1 - .3636)^2 = .5950$

d. We want the 10th percentile: $.10 = F(x; 9, 180) = 1 - e^{-\left(\frac{x}{180}\right)^9}$. A bit of algebra leads us to $x = 140.178$. Thus 10% of all tensile strengths will be less than 140.178 MPa.

113.

a. $F_Y(y) = P(Y \le y) = P(60X \le y) = P\left(X \le \dfrac{y}{60}\right) = F\left(\dfrac{y}{60\beta}; \alpha\right).$ Thus $f_Y(y)$

$$= f\left(\dfrac{y}{60\beta}; \alpha\right) \cdot \dfrac{1}{60\beta} = \dfrac{y^{\alpha-1} e^{\frac{-y}{60\beta}}}{(60\beta)^\alpha \Gamma(\alpha)},$$ which shows that Y has a gamma

distribution with parameters α and 60β.

b. With c replacing 60 in **a**, the same argument shows that cX has a gamma distribution with parameters α and $c\beta$.

115.

a. Assuming independence, P(all 3 births occur on March 11) $= \left(\frac{1}{365}\right)^3 = .00000002$

b. $\left(\frac{1}{365}\right)^3 (365) = .0000073$

c. Let X = deviation from due date. X~N(0, 19.88). Then the baby due on March 15 was 4 days early. $P(x = -4) \approx P(-4.5 < x < -3.5)$

$$= \Phi\left(\dfrac{-3.5}{19.88}\right) - \Phi\left(\dfrac{-4.5}{19.88}\right) = \Phi(-.18) - \Phi(-.237) = .4286 - .4090 = .0196.$$

Similarly, the baby due on April 1 was 21 days early, and $P(x = -21)$

$$\approx \Phi\left(\dfrac{-20.5}{19.88}\right) - \Phi\left(\dfrac{-21.5}{19.88}\right) = \Phi(-1.03) - \Phi(-1.08) = .1515 - .1401 = .0114.$$

The baby due on April 4 was 24 days early, and $P(x = -24) \approx .0097$

Again, assuming independence, P(all 3 births occurred on March 11) =
$(.0196)(.0114)(.0097) = .00002145$

d. To calculate the probability of the three births happening on any day, we could make similar calculations as in part c for each possible day, and then add the probabilities.

117.

a. $F_X(x) = P\left(-\dfrac{1}{\lambda}\ln(1-U) \le x\right) = P\big(\ln(1-U) \ge -\lambda x\big) = P\big(1-U \ge e^{-\lambda x}\big)$

$= P\big(U \le 1-e^{-\lambda x}\big) = 1-e^{-\lambda x}$ since $F_U(u) = u$ (U is uniform on [0, 1]). Thus X has an exponential distribution with parameter λ.

b. By taking successive random numbers u_1, u_2, u_3, …and computing

$x_i = -\dfrac{1}{10}\ln(1-u_i)$, … we obtain a sequence of values generated from an

exponential distribution with parameter $\lambda = 10$.

119. $g(\mu) + g'(\mu)(X - \mu) \le g(X)$ implies that $E[g(\mu) + g'(\mu)(X - \mu)] = E(g(\mu)) = g(\mu) \le E(g(X))$, i.e. that $g(E(X)) \le E(g(X))$.

CHAPTER 5

Section 5.1

1.

   a. $P(X = 1, Y = 1) = p(1,1) = .20$

   b. $P(X \leq 1 \text{ and } Y \leq 1) = p(0,0) + p(0,1) + p(1,0) + p(1,1) = .42$

   c. At least one hose is in use at both islands. $P(X \neq 0 \text{ and } Y \neq 0) = p(1,1) + p(1,2) + p(2,1) + p(2,2) = .70$

   d. By summing row probabilities, $p_x(x) = .16, .34, .50$ for $x = 0, 1, 2$, and by summing column probabilities, $p_y(y) = .24, .38, .38$ for $y = 0, 1, 2$. $P(X \leq 1) = p_x(0) + p_x(1) = .50$

   e. $P(0,0) = .10$, but $p_x(0) \cdot p_y(0) = (.16)(.24) = .0384 \neq .10$, so X and Y are not independent.

3.

   a. $p(1,1) = .15$, the entry in the 1st row and 1st column of the joint probability table.

   b. $P(X_1 = X_2) = p(0,0) + p(1,1) + p(2,2) + p(3,3) = .08 + .15 + .10 + .07 = .40$

   c. $A = \{ (x_1, x_2): x_1 \geq 2 + x_2 \} \cup \{ (x_1, x_2): x_2 \geq 2 + x_1 \}$
      $P(A) = p(2,0) + p(3,0) + p(4,0) + p(3,1) + p(4,1) + p(4,2) + p(0,2) + p(0,3) + p(1,3) = .22$

   d. $P(\text{exactly } 4) = p(1,3) + p(2,2) + p(3,1) + p(4,0) = .17$
      $P(\text{at least } 4) = P(\text{exactly } 4) + p(4,1) + p(4,2) + p(4,3) + p(3,2) + p(3,3) + p(2,3) = .46$

5.

a. $P(X = 3, Y = 3) = P(3 \text{ customers, each with 1 package})$
  $= P(\text{ each has 1 package} \mid 3 \text{ customers}) \cdot P(3 \text{ customers})$
  $= (.6)^3 \cdot (.25) = .054$

b. $P(X = 4, Y = 11) = P(\text{total of 11 packages} \mid 4 \text{ customers}) \cdot P(4 \text{ customers})$
  Given that there are 4 customers, there are 4 different ways to have a total of 11
  packages: 3, 3, 3, 2 or 3, 3, 2, 3 or 3, 2, 3 ,3 or 2, 3, 3, 3. Each way has
  probability $(.1)^3(.3)$, so $p(4, 11) = 4(.1)^3(.3)(.15) = .00018$

c. $p(x,y) = 0$ unless $y = 0, 1, \ldots, x$; $x = 0, 1, 2, 3, 4$. For any such pair,

$$p(x,y) = P(Y = y \mid X = x) \cdot P(X = x) = \binom{x}{y}(.6)^y (.4)^{x-y} \cdot p_x(x)$$

$$p_y(4) = p(y = 4) = p(x = 4, y = 4) = p(4,4) = (.6)^4 \cdot (.15) = .0194$$

$$p_y(3) = p(3,3) + p(4,3) = (.6)^3(.25) + \binom{4}{3}(.6)^3(.4)(.15) = .1058$$

$$p_y(2) = p(2,2) + p(3,2) + p(4,2) = (.6)^2(.3) + \binom{3}{2}(.6)^2(.4)(.25)$$

$$+ \binom{4}{2}(.6)^2(.4)^2(.15) = .2678$$

$$p_y(1) = p(1,1) + p(2,1) + p(3,1) + p(4,1) = (.6)(.2) + \binom{2}{1}(.6)(.4)(.3)$$

$$\binom{3}{1}(.6)(.4)^2(.25) + \binom{4}{1}(.6)(.4)^3(.15) = .3590$$

$p_y(0) = 1 - [.3590 + .2678 + .1058 + .0194] = .2480$

7.

  a. $p(1,1) = .030$

  b. $P(X \le 1 \text{ and } Y \le 1 = p(0,0) + p(0,1) + p(1,0) + p(1,1) = .120$

  c. $P(X = 1) = p(1,0) + p(1,1) + p(1,2) = .100; P(Y = 1) = p(0,1) + \ldots + p(5,1) = .300$

  d. $P(\text{overflow}) = P(X + 3Y > 5) = 1 - P(X + 3Y \le 5) = 1 - P[(X,Y) = (0,0) \text{ or } \ldots \text{or}$
  $(5,0) \text{ or } (0,1) \text{ or } (1,1) \text{ or } (2,1)] = 1 - .620 = .380$

  e. The marginal probabilities for X (row sums from the joint probability table) are
  $p_x(0) = .05, p_x(1) = .10, p_x(2) = .25, p_x(3) = .30, p_x(4) = .20, p_x(5) = .10$; those
  for Y (column sums) are $p_y(0) = .5, p_y(1) = .3, p_y(2) = .2$. It is now easily verified
  that for every (x,y), $p(x,y) = p_x(x) \cdot p_y(y)$, so X and Y are independent.

9.

  a. $1 = \int_{-\infty}^{\infty} \int_{-\infty}^{\infty} f(x, y) dx dy = \int_{20}^{30} \int_{20}^{30} K(x^2 + y^2) dx dy$

  $= K \int_{20}^{30} \int_{20}^{30} x^2 dy dx + K \int_{20}^{30} \int_{20}^{30} y^2 dx dy = 10K \int_{20}^{30} x^2 dx + 10K \int_{20}^{30} y^2 dy$

  $= 20K \cdot \left(\frac{19,000}{3} \right) \Rightarrow K = \frac{3}{380,000}$

  b. $P(X < 26 \text{ and } Y < 26) = \int_{20}^{26} \int_{20}^{26} K(x^2 + y^2) dx dy = 12K \int_{20}^{26} x^2 dx$

  $4Kx^3 \Big|_{20}^{26} = 38,304K = .3024$

c.

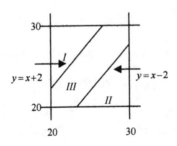

$$P(\,|\,X-Y\,|\le 2\,) = \iint_{\substack{region \\ III}} f(x,y)\,dxdy$$

$$1 - \iint_{I} f(x,y)\,dxdy - \iint_{II} f(x,y)\,dxdy$$

$$1 - \int_{20}^{28}\int_{x+2}^{30} f(x,y)\,dydx - \int_{22}^{30}\int_{20}^{x-2} f(x,y)\,dydx$$

= (after much algebra) .3593

d. $f_x(x) = \int_{-\infty}^{\infty} f(x,y)\,dy = \int_{20}^{30} K(x^2 + y^2)\,dy = 10Kx^2 + K\frac{y^3}{3}\Big|_{20}^{30}$

$= 10Kx^2 + .05, \qquad\qquad 20 \le x \le 30$

e. $f_y(y)$ is obtained by substituting y for x in (d); clearly $f(x,y) \ne f_x(x) \cdot f_y(y)$, so X and Y are not independent.

11.

a. $p(x,y) = \dfrac{e^{-\lambda}\lambda^x}{x!} \cdot \dfrac{e^{-\mu}\mu^y}{y!}$ for x = 0, 1, 2, ...; y = 0, 1, 2, ...

b. $p(0,0) + p(0,1) + p(1,0) = e^{-\lambda-\mu}\left[1 + \lambda + \mu\right]$

c. $P(X+Y=m) = \displaystyle\sum_{k=0}^{m} P(X = k, Y = m-k) = \sum_{k=0}^{m} e^{-\lambda-\mu}\dfrac{\lambda^k}{k!}\dfrac{\mu^{m=k}}{(m-k)!}$

$\dfrac{e^{-(\lambda+\mu)}}{m!}\displaystyle\sum_{k=0}^{m}\binom{m}{k}\lambda^k\mu^{m-k} = \dfrac{e^{-(\lambda+\mu)}(\lambda+\mu)^m}{m!}$, so the total # of errors X+Y

also has a Poisson distribution with parameter $\lambda + \mu$.

13.

a. $f(x,y) = f_x(x) \cdot f_y(y) = \begin{cases} e^{-x-y} & x \ge 0, y \ge 0 \\ 0 & otherwise \end{cases}$

b. $P(X \le 1 \text{ and } Y \le 1) = P(X \le 1) \cdot P(Y \le 1) = (1 - e^{-1})(1 - e^{-1}) = .400$

c. $P(X + Y \le 2) = \displaystyle\int_0^2\int_0^{2-x} e^{-x-y}\,dydx = \int_0^2 e^{-x}\left[1 - e^{-(2-x)}\right]dx$

$= \displaystyle\int_0^2 (e^{-x} - e^{-2})dx = 1 - e^{-2} - 2e^{-2} = .594$

d. $P(X + Y \le 1) = \displaystyle\int_0^1 e^{-x}\left[1 - e^{-(1-x)}\right]dx = 1 - 2e^{-1} = .264$,

so $P(1 \le X + Y \le 2) = P(X + Y \le 2) - P(X + Y \le 1) = .594 - .264 = .330$

15.

a. $F(y) = P(Y \le y) = P[(X_1 \le y) \cup ((X_2 \le y) \cap (X_3 \le y))]$
$= P(X_1 \le y) + P[(X_2 \le y) \cap (X_3 \le y)] - P[(X_1 \le y) \cap (X_2 \le y) \cap (X_3 \le y)]$
$= (1 - e^{-\lambda y}) + (1 - e^{-\lambda y})^2 - (1 - e^{-\lambda y})^3$ for $y \ge 0$

b. $f(y) = F'(y) = \lambda e^{-\lambda y} + 2(1 - e^{-\lambda y})(\lambda e^{-\lambda y}) - 3(1 - e^{-\lambda y})^2 (\lambda e^{-\lambda y})$
$= 4\lambda e^{-2\lambda y} - 3\lambda e^{-3\lambda y}$ for $y \ge 0$

c. $E(Y) = \int_0^\infty y \cdot (4\lambda e^{-2\lambda y} - 3\lambda e^{-3\lambda y}) dy = 2\left(\frac{1}{2\lambda}\right) - \frac{1}{3\lambda} = \frac{2}{3\lambda}$

17.

a. $P((X,Y) \text{ within a circle of radius } \frac{R}{2}) = P(A) = \iint_A f(x,y)dxdy$

$= \frac{1}{\pi R^2} \iint_A dxdy = \frac{area.of.A}{\pi R^2} = \frac{1}{4} = .25$

b.

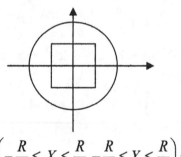

$P\left(-\frac{R}{2} \le X \le \frac{R}{2}, -\frac{R}{2} \le Y \le \frac{R}{2}\right) = \frac{R^2}{\pi R^2} = \frac{1}{\pi}$

c.

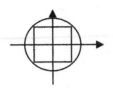

$$P\left(-\frac{R}{\sqrt{2}} \le X \le \frac{R}{\sqrt{2}}, -\frac{R}{\sqrt{2}} \le Y \le \frac{R}{\sqrt{2}}\right) = \frac{2R^2}{\pi R^2} = \frac{2}{\pi}$$

d. $f_x(x) = \int_{-\infty}^{\infty} f(x,y)dy = \int_{-\sqrt{R^2-x^2}}^{\sqrt{R^2-x^2}} \frac{1}{\pi R^2} dy = \frac{2\sqrt{R^2-x^2}}{\pi R^2}$ for $-R \le x \le R$

and similarly for $f_Y(y)$. X and Y are not independent since e.g. $f_x(.9R) = f_Y(.9R)$ > 0, yet f(.9R, .9R) = 0 since (.9R, .9R) is outside the circle of radius R.

19.

a. $f_{Y|X}(y\,|\,x) = \frac{f(x,y)}{f_X(x)} = \frac{k(x^2 + y^2)}{10kx^2 + .05}$ $20 \le y \le 30$

$f_{X|Y}(x\,|\,y) = \frac{k(x^2 + y^2)}{10ky^2 + .05}$ $20 \le x \le 30$ $\left(k = \frac{3}{380,000}\right)$

b. $P(\,Y \ge 25 \,|\, X = 22\,) = \int_{25}^{30} f_{Y|X}(y\,|\,22)dy$

$= \int_{25}^{30} \frac{k((22)^2 + y^2)}{10k(22)^2 + .05} dy = .783$

$P(\,Y \ge 25\,) = \int_{25}^{30} f_Y(y)dy = \int_{25}^{30}(10ky^2 + .05)dy = .75$

c. $E(\,Y\,|\,X{=}22\,) = \int_{-\infty}^{\infty} y \cdot f_{Y|X}(y\,|\,22)dy = \int_{20}^{30} y \cdot \frac{k((22)^2 + y^2)}{10k(22)^2 + .05} dy$

$= 25.372912$

$E(\,Y^2\,|\,X{=}22\,) = \int_{20}^{30} y^2 \cdot \frac{k((22)^2 + y^2)}{10k(22)^2 + .05} dy = 652.028640$

$V(Y|\,X = 22\,) = E(\,Y^2\,|\,X{=}22\,) - [E(\,Y\,|\,X{=}22\,)]^2 = 8.243976$

21. For every x and y, $f_{Y|X}(y|x) = f_y(y)$, since then $f(x,y) = f_{Y|X}(y|x) \cdot f_X(x) = f_Y(y) \cdot f_X(x)$, as required.

Section 5.2

23. $E(X_1 - X_2) = \displaystyle\sum_{x_1=0}^{4} \sum_{x_2=0}^{3} (x_1 - x_2) \cdot p(x_1, x_2) =$

$(0 - 0)(.08) + (0 - 1)(.07) + \ldots + (4 - 3)(.06) = .15$
(which also equals $E(X_1) - E(X_2) = 1.70 - 1.55$)

25. $E(XY) = E(X) \cdot E(Y) = L \cdot L = L^2$

27. $E[h(X,Y)] = \displaystyle\int_0^1 \int_0^1 |x - y| \cdot 6x^2 y \, dx \, dy = 2 \int_0^1 \int_0^x (x - y) \cdot 6x^2 y \, dy \, dx$

$12 \displaystyle\int_0^1 \int_0^x (x^3 y - x^2 y^2) \, dy \, dx = 12 \int_0^1 \frac{x^5}{6} \, dx = \frac{1}{3}$

29. $\text{Cov}(X,Y) = -\dfrac{2}{75}$ and $\mu_x = \mu_y = \dfrac{2}{5}$. $E(X^2) = \displaystyle\int_0^1 x^2 \cdot f_x(x) \, dx$

$= 12 \displaystyle\int_0^1 x^3 (1 - x^2 \, dx) = \frac{12}{60} = \frac{1}{5}$, so Var (X) $= \dfrac{1}{5} - \dfrac{4}{25} = \dfrac{1}{25}$

Similarly, Var(Y) $= \dfrac{1}{25}$, so $\rho_{X,Y} = \dfrac{-\frac{2}{75}}{\sqrt{\frac{1}{25}} \cdot \sqrt{\frac{1}{25}}} = -\dfrac{50}{75} = -.667$

Chapter 5: Joint Probability Distributions and Random Samples

31.

a. $E(X) = \int_{20}^{30} x f_x(x)dx = \int_{20}^{30} x \left[10Kx^2 + .05\right]dx = 25.329 = E(Y)$

$E(XY) = \int_{20}^{30} \int_{20}^{30} xy \cdot K(x^2 + y^2)dxdy = 641.447$

$\Rightarrow Cov(X,Y) = 641.447 - (25.329)^2 = -.111$

b. $E(X^2) = \int_{20}^{30} x^2 \left[10Kx^2 + .05\right]dx = 649.8246 = E(Y^2)$,

so Var (X) = Var(Y) = 649.8246 – $(25.329)^2$ = 8.2664

$\Rightarrow \rho = \dfrac{-.111}{\sqrt{(8.2664)(8.2664)}} = -.0134$

33. Since E(XY) = E(X) · E(Y), Cov(X,Y) = E(XY) – E(X) · E(Y) = E(X) · E(Y) - E(X) ·

E(Y) = 0, and since Corr(X,Y) = $\dfrac{Cov(X,Y)}{\sigma_x \sigma_y}$, then Corr(X,Y) = 0

35.

a. Cov(aX + b, cY + d) = E[(aX + b)(cY + d)] – E(aX + b) · E(cY + d)
= E[acXY + adX + bcY + bd] – (aE(X) + b)(cE(Y) + d)
= acE(XY) – acE(X)E(Y) = acCov(X,Y)

b. Corr(aX + b, cY + d) =

$\dfrac{Cov(aX + b, cY + d)}{\sqrt{Var(aX + b)}\sqrt{Var(cY + d)}} = \dfrac{acCov(X,Y)}{|a| \cdot |c| \sqrt{Var(X) \cdot Var(Y)}}$

= Corr(X,Y) when a and c have the same signs. When a and c differ in sign,
Corr(aX + b, cY + d) = -Corr(X,Y).

Section 5.3

37.

P(x₁)	.20	.50	.30

P(x₂)	$x_2 \mid x_1$	25	40	65
.20	25	.04	.10	.06
.50	40	.10	.25	.15
.30	65	.06	.15	.09

a.

$\bar{x}$	25	32.5	40	45	52.5	65
$p(\bar{x})$	.04	.20	.25	.12	.30	.09

$$E(\bar{x}) = (25)(.04) + 32.5(.20) + ... + 65(.09) = 44.5 = \mu$$

b.

s^2	0	112.5	312.5	800
$P(s^2)$	.38	.20	.30	.12

$$E(s^2) = 212.25 = \sigma^2$$

39.

x	0	1	2	3	4	5	6	7	8	9	10
x/n	0	.1	.2	.3	.4	.5	.6	.7	.8	.9	1.0
p(x/n)	.000	.000	.000	.001	.005	.027	.088	.201	.302	.269	.107

X is a binomial random variable with p = .8.

41.

Outcome	1,1	1,2	1,3	1,4	2,1	2,2	2,3	2,4
Probability	.16	.12	.08	.04	.12	.09	.06	.03
$\bar{x}$	1	1.5	2	2.5	1.5	2	2.5	3
r	0	1	2	3	1	0	1	2

Outcome	3,1	3,2	3,3	3,4	4,1	4,2	4,3	4,4
Probability	.08	.06	.04	.02	.04	.03	.02	.01
$\bar{x}$	2	2.5	3	3.5	2.5	3	3.5	4
r	2	1	0	1	3	2	1	2

a.

$\bar{x}$	1	1.5	2	2.5	3	3.5	4
$p(\bar{x})$	.16	.24	.25	.20	.10	.04	.01

b. $P(\bar{x} \le 2.5) = .8$

c.

r	0	1	2	3
p(r)	.30	.40	.22	.08

d. $P(\overline{X} \le 1.5) = P(1,1,1,1) + P(2,1,1,1) + \ldots + P(1,1,1,2) + P(1,1,2,2) + \ldots +$
P(2,2,1,1) + P(3,1,1,1) + \ldots + P(1,1,1,3)
$= (.4)^4 + 4(.4)^3(.3) + 6(.4)^2(.3)^2 + 4(.4)^2(.2)^2 = .2400$

43.   The statistic of interest is the fourth spread, or the difference between the medians of the upper and lower halves of the data. The population distribution is uniform with A = 8 and B = 10.   Use a computer to generate samples of sizes n = 5, 10, 20, and 30 from a uniform distribution with A = 8 and B = 10. Keep the number of replications the same (say 500, for example). For each sample, compute the upper and lower fourth, then compute the difference. Plot the sampling distributions on separate histograms for n = 5, 10, 20, and 30.

45. Using Minitab to generate the necessary sampling distribution, we can see that as n increases, the distribution slowly moves toward normality. However, even the sampling distribution for n = 50 is not yet approximately normal.
n = 10

n = 50

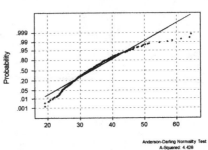

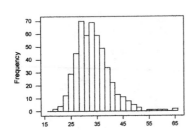

Section 5.4

47. $\mu = 12$ cm $\qquad \sigma = .04$ cm

a. n = 16 $\quad P(11.99 \le \overline{X} \le 12.01) = P\left(\dfrac{11.99 - 12}{.01} \le Z \le \dfrac{12.01 - 12}{.01} \right)$

$$= P(-1 \le Z \le 1)$$
$$= \Phi(1) - \Phi(-1)$$
$$= .8413 - .1587$$
$$= .6826$$

b. n = 25 $\quad P(\overline{X} > 12.01) = P\left(Z > \dfrac{12.01 - 12}{.04 / 5} \right) = P(Z > 1.25)$

$$= 1 - \Phi(1.25)$$
$$= 1 - .8944$$
$$= .1056$$

49.

    a. 11 P.M. – 6:50 P.M. = 250 minutes. With $T_0 = X_1 + \ldots + X_{40}$ = total grading time, $\mu_{T_0} = n\mu = (40)(6) = 240$ and $\sigma_{T_0} = \sigma\sqrt{n} = 37.95$, so P($T_0 \le$

$$250) \approx P\left(Z \le \frac{250 - 240}{37.95}\right) = P(Z \le .26) = .6026$$

    b. $P(T_0 > 260) = P\left(Z > \dfrac{260 - 240}{37.95}\right) = P(Z > .53) = .2981$

51. $X \sim N(10,4)$. For day 1, n = 5

$$P(\overline{X} \le 11) = P\left(Z \le \frac{11 - 10}{2/\sqrt{5}}\right) = P(Z \le 1.12) = .8686$$

For day 2, n = 6

$$P(\overline{X} \le 11) = P\left(Z \le \frac{11 - 10}{2/\sqrt{6}}\right) = P(Z \le 1.22) = .8888$$

For both days,

$P(\overline{X} \le 11) = (.8686)(.8888) = .7720$

53. $\mu = 50$, $\sigma = 1.2$

    a. n = 9

$$P(\overline{X} \ge 51) = P\left(Z \ge \frac{51 - 50}{1.2/\sqrt{9}}\right) = P(Z \ge 2.5) = 1 - .9938 = .0062$$

    b. n = 40

$$P(\overline{X} \ge 51) = P\left(Z \ge \frac{51 - 50}{1.2/\sqrt{40}}\right) = P(Z \ge 5.27) \approx 0$$

55. $\mu = np = 20 \quad \sigma = \sqrt{npq} = 3.464$

    a. $P(25 \le X) \approx P\left(\dfrac{24.5 - 20}{3.464} \le Z\right) = P(1.30 \le Z) = .0968$

    b. $P(15 \le X \le 25) \approx P\left(\dfrac{14.5 - 20}{3.464} \le Z \le \dfrac{25.5 - 20}{3.464}\right)$

        $= P(-1.59 \le Z \le 1.59) = .8882$

57. $E(X) = 100$, $Var(X) = 200$, $\sigma_x = 14.14$, so $P(X \le 125) \approx P\left(Z \le \dfrac{125 - 100}{14.14}\right)$

    $= P(Z \le 1.77) = .9616$

Section 5.5

59.

    a. $E(X_1 + X_2 + X_3) = 180$, $V(X_1 + X_2 + X_3) = 45$, $\sigma_{x_1 + x_2 + x_3} = 6.708$

        $P(X_1 + X_2 + X_3 \le 200) = P\left(Z \le \dfrac{200 - 180}{6.708}\right) = P(Z \le 2.98) = .9986$

        $P(150 \le X_1 + X_2 + X_3 \le 200) = P(-4.47 \le Z \le 2.98) \approx .9986$

    b. $\mu_{\bar{X}} = \mu = 60$, $\sigma_{\bar{x}} = \dfrac{\sigma_x}{\sqrt{n}} = \dfrac{\sqrt{15}}{\sqrt{3}} = 2.236$

        $P(\bar{X} \ge 55) = P\left(Z \ge \dfrac{55 - 60}{2.236}\right) = P(Z \ge -2.236) = .9875$

        $P(58 \le \bar{X} \le 62) = P(-.89 \le Z \le .89) = .6266$

c. $E(X_1 - .5X_2 - .5X_3) = 0;$

   $V(X_1 - .5X_2 - .5X_3) = \sigma_1^2 + .25\sigma_2^2 + .25\sigma_3^2 = 22.5$, sd = 4.7434

   $P(-10 \leq X_1 - .5X_2 - .5X_3 \leq 5) = P\left(\dfrac{-10-0}{4.7434} \leq Z \leq \dfrac{5-0}{4.7434} \right)$

   $= P\left(-2.11 \leq Z \leq 1.05 \right) = .8531 - .0174 = .8357$

d. $E(X_1 + X_2 + X_3) = 150,\ V(X_1 + X_2 + X_3) = 36,\ \sigma_{x_1+x_2+x_3} = 6$

   $P(X_1 + X_2 + X_3 \leq 200) = P\left(Z \leq \dfrac{160-150}{6} \right) = P(Z \leq 1.67) = .9525$

   We want $P(X_1 + X_2 \geq 2X_3)$, or written another way, $P(X_1 + X_2 - 2X_3 \geq 0)$
   $E(X_1 + X_2 - 2X_3) = 40 + 50 - 2(60) = -30,$
   $V(X_1 + X_2 - 2X_3) = \sigma_1^2 + \sigma_2^2 + 4\sigma_3^2 = 78, 36$, sd = 8.832, so

   $P(X_1 + X_2 - 2X_3 \geq 0) = P\left(Z \geq \dfrac{0-(-30)}{8.832} \right) = P(Z \geq 3.40) = .0003$

61.

a. The marginal pmf's of X and Y are given in the solution to Exercise 7, from which $E(X) = 2.8, E(Y) = .7, V(X) = 1.66, V(Y) = .61$. Thus $E(X+Y) = E(X) + E(Y) = 3.5, V(X+Y) = V(X) + V(Y) = 2.27$, and the standard deviation of $X + Y$ is 1.51

b. $E(3X+10Y) = 3E(X) + 10E(Y) = 15.4, V(3X+10Y) = 9V(X) + 100V(Y) = 75.94$, and the standard deviation of revenue is 8.71

63.

a. $E(X_1) = 1.70, E(X_2) = 1.55, E(X_1X_2) = \displaystyle\sum_{x_1}\sum_{x_2} x_1x_2 p(x_1, x_2) = 3.33$, so

   $Cov(X_1,X_2) = E(X_1X_2) - E(X_1) E(X_2) = 3.33 - 2.635 = .695$

b. $V(X_1 + X_2) = V(X_1) + V(X_2) + 2\ Cov(X_1,X_2)$
   $= 1.59 + 1.0875 + 2(.695) = 4.0675$

65. $\mu = 5.00$, $\sigma = .2$

   a. $E(\overline{X} - \overline{Y}) = 0;$ $V(\overline{X} - \overline{Y}) = \dfrac{\sigma^2}{25} + \dfrac{\sigma^2}{25} = .0032$, $\sigma_{\overline{X}-\overline{Y}} = .0566$

   $\Rightarrow P\left(-.1 \le \overline{X} - \overline{Y} \le .1\right) \approx P\left(-1.77 \le Z \le 1.77\right) = .9232$ (by the CLT)

   b. $V(\overline{X} - \overline{Y}) = \dfrac{\sigma^2}{36} + \dfrac{\sigma^2}{36} = .0022222$, $\sigma_{\overline{X}-\overline{Y}} = .0471$

   $\Rightarrow P\left(-.1 \le \overline{X} - \overline{Y} \le .1\right) \approx P\left(-2.12 \le Z \le 2.12\right) = .9660$

67.   Letting X_1, X_2, and X_3 denote the lengths of the three pieces, the total length is
$X_1 + X_2 - X_3$. This has a normal distribution with mean value $20 + 15 - 1 = 34$,
variance $.25 + .16 + .01 = .42$, and standard deviation $.6481$. Standardizing gives
$P(34.5 \le X_1 + X_2 - X_3 \le 35) = P(.77 \le Z \le 1.54) = .1588$

69.

   a. $E(X_1 + X_2 + X_3) = 800 + 1000 + 600 = 2400$.

   b.   Assuming independence of X_1, X_2, X_3, $\text{Var}(X_1 + X_2 + X_3)$
$= (16)^2 + (25)^2 + (18)^2 = 12.05$

   c. $E(X_1 + X_2 + X_3) = 2400$ as before, but now $\text{Var}(X_1 + X_2 + X_3)$
$= \text{Var}(X_1) + \text{Var}(X_2) + \text{Var}(X_3) + 2\text{Cov}(X_1,X_2) + 2\text{Cov}(X_1, X_3) + 2\text{Cov}(X_2, X_3) =$
1745, with sd $= 41.77$

71.

   a. $M = a_1 X_1 + a_2 X_2 + W \displaystyle\int_0^{12} x\,dx = a_1 X_1 + a_2 X_2 + 72W$, so
$E(M) = (5)(2) + (10)(4) + (72)(1.5) = 158\text{m}$
$\sigma_M^2 = (5)^2(.5)^2 + (10)^2(1)^2 + (72)^2(.25)^2 = 430.25$, $\sigma_M = 20.74$

   b. $P(M \le 200) = P\left(Z \le \dfrac{200-158}{20.74}\right) = P(Z \le 2.03) = .9788$

Chapter 5: Joint Probability Distributions and Random Samples

73.

    a.   Both approximately normal by the C.L.T.

    b.   The difference of two r.v.'s is just a special linear combination, and a linear combination of normal r.v's has a normal distribution, so $\overline{X} - \overline{Y}$ has approximately a normal distribution with $\mu_{\overline{X}-\overline{Y}} = 5$ and

$$\sigma^2_{\overline{X}-\overline{Y}} = \frac{8^2}{40} + \frac{6^2}{35} = 2.629, \sigma_{\overline{X}-\overline{Y}} = 1.621$$

    c. $P\left(-1 \le \overline{X} - \overline{Y} \le 1\right) \doteq P\left(\dfrac{-1-5}{1.6213} \le Z \le \dfrac{1-5}{1.6213}\right)$

             $= P(-3.70 \le Z \le -2.47) \approx .0068$

    d. $P\left(\overline{X} - \overline{Y} \ge 10\right) \doteq P\left(Z \ge \dfrac{10-5}{1.6213}\right) = P(Z \ge 3.08) = .0010.$ This

       probability is quite small, so such an occurrence is unlikely if $\mu_1 - \mu_2 = 5$, and we would thus doubt this claim.

Supplementary Exercises

75.

    a. $p_X(x)$ is obtained by adding joint probabilities across the row labeled x, resulting in $p_X(x) = .2, .5, .3$ for $x = 12, 15, 20$ respectively. Similarly, from column sums $p_y(y) = .1, .35, .55$ for $y = 12, 15, 20$ respectively.

    b. $P(X \le 15$ and $Y \le 15) = p(12,12) + p(12,15) + p(15,12) + p(15,15) = .25$

    c. $p_X(12) \cdot p_y(12) = (.2)(.1) \ne .05 = p(12,12)$, so X and Y are not independent. (Almost any other (x,y) pair yields the same conclusion).

    d. $E(X + Y) = \sum\sum(x+y)p(x,y) = 33.35$ (or $= E(X) + E(Y) = 33.35$)

    e. $E(|X - Y|) = \sum\sum|x+y|p(x,y) = 3.85$

77.

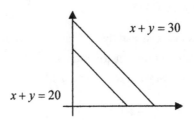

a. $1 = \int_{-\infty}^{\infty} \int_{-\infty}^{\infty} f(x,y)dxdy = \int_0^{20} \int_{20-x}^{30-x} kxydydx + \int_{20}^{30} \int_0^{30-x} kxydydx$

$= \dfrac{81{,}250}{3} \cdot k \Rightarrow k = \dfrac{3}{81{,}250}$

b. $f_X(x) = \begin{cases} \int_{20-x}^{30-x} kxydy = k(250x - 10x^2) & 0 \le x \le 20 \\ \int_0^{30-x} kxydy = k(450x - 30x^2 + \frac{1}{2}x^3) & 20 \le x \le 30 \end{cases}$

and by symmetry $f_Y(y)$ is obtained by substituting y for x in $f_X(x)$. Since $f_X(25) >$ 0, and $f_Y(25) > 0$, but $f(25, 25) = 0$, $f_X(x) \cdot f_Y(y) \ne f(x,y)$ for all x,y so X and Y are not independent.

c. $P(X + Y \le 25) = \int_0^{20} \int_{20-x}^{25-x} kxydydx + \int_{20}^{25} \int_0^{25-x} kxydydx$

$= \dfrac{3}{81{,}250} \cdot \dfrac{230{,}625}{24} = .355$

d. $E(X + Y) = E(X) + E(Y) = 2\Big\{ \int_0^{20} x \cdot k\big(250x - 10x^2\big)dx$

$+ \int_{20}^{30} x \cdot k\big(450x - 30x^2 + \frac{1}{2}x^3\big)dx \Big\} \quad = 2k(351{,}666.67) = 25.969$

e. $E(XY) = \int_{-\infty}^{\infty} \int_{-\infty}^{\infty} xy \cdot f(x,y)dxdy = \int_{0}^{20} \int_{20-x}^{30-x} kx^2 y^2 dydx$

$\qquad + \int_{20}^{30} \int_{0}^{30-x} kx^2 y^2 dydx = \frac{k}{3} \cdot \frac{33,250,000}{3} = 136.4103$, so

Cov(X,Y) = $136.4103 - (12.9845)^2 = -32.19$, and $E(X^2) = E(Y^2) = 204.6154$, so

$\sigma_x^2 = \sigma_y^2 = 204.6154 - (12.9845)^2 = 36.0182$ and

$\rho = \dfrac{-32.19}{36.0182} = -.894$

f. Var (X + Y) = Var(X) + Var(Y) + 2Cov(X,Y) = 7.66

79. $E(\overline{X} + \overline{Y} + \overline{Z}) = 500 + 900 + 2000 = 3400$

$Var(\overline{X} + \overline{Y} + \overline{Z}) = \dfrac{50^2}{365} + \dfrac{100^2}{365} + \dfrac{180^2}{365} = 123.014$, and the std dev = 11.09.

$P(\overline{X} + \overline{Y} + \overline{Z} \le 3500) = P(Z \le 9.0) \approx 1$

81.

a. $E(N) \cdot \mu = (10)(40) = 400$ minutes

b. We expect 20 components to come in for repair during a 4 hour period,
so $E(N) \cdot \mu = (20)(3.5) = 70$

83. $0.95 = P(\mu - .02 \le \overline{X} \le \mu + .02) \doteq P\left(\dfrac{-.02}{.01/\sqrt{n}} \le Z \le \dfrac{.02}{.01/\sqrt{n}} \right)$

$\qquad = P\left(-.2\sqrt{n} \le Z \le .2\sqrt{n} \right)$, but $P(-1.96 \le Z \le 1.96) = .95$ so
$.2\sqrt{n} = 1.96 \Rightarrow n = 97$. The C.L.T.

85. The expected value and standard deviation of volume are 87,850 and 4370.37, respectively, so

$P(volume \le 100,000) = P\left(Z \le \dfrac{100,000 - 87,850}{4370.37} \right) = P(Z \le 2.78) = .9973$

87.

a. $Var(aX + Y) = a^2\sigma_x^2 + 2aCov(X,Y) + \sigma_y^2 = a^2\sigma_x^2 + 2a\sigma_X\sigma_Y\rho + \sigma_y^2$.

Substituting $a = \dfrac{\sigma_Y}{\sigma_X}$ yields $\sigma_Y^2 + 2\sigma_Y^2\rho + \sigma_Y^2 = 2\sigma_Y^2(1-\rho) \geq 0$, so

$\rho \geq -1$

b. Same argument as in **a**

c. Suppose $\rho = 1$. Then $Var(aX - Y) = 2\sigma_Y^2(1-\rho) = 0$, which implies that $aX - Y = k$ (a constant), so $aX - Y = aX - k$, which is of the form $aX + b$.

89.

a. With $Y = X_1 + X_2$,

$$F_Y(y) = \int_0^y \left\{ \int_0^{y-x_1} \frac{1}{2^{v_1/2}\Gamma(v_1/2)} \cdot \frac{1}{2^{v_{21}/2}\Gamma(v_2/2)} \cdot x_1^{\frac{v_1}{2}-1} x_2^{\frac{v_2}{2}-1} e^{-\frac{x_1+x_2}{2}} dx_2 \right\} dx_1$$

. But the inner integral can be shown to be equal to

$\dfrac{1}{2^{(v_1+v_2)/2}\Gamma((v_1+v_2)/2)} y^{[(v_1+v_2)/2]-1} e^{-y/2}$, from which the result follows.

b. By **a**, $Z_1^2 + Z_2^2$ is chi-squared with $v = 2$, so $\left(Z_1^2 + Z_2^2\right) + Z_3^2$ is chi-squared with $v = 3$, etc, until $Z_1^2 + ... + Z_n^2$ 9s chi-squared with $v = n$

c. $\dfrac{X_i - \mu}{\sigma}$ is standard normal, so $\left[\dfrac{X_i - \mu}{\sigma}\right]^2$ is chi-squared with $v = 1$, so the sum is chi-squared with $v = n$.

91.

a. $V(X_1) = V(W + E_1) = \sigma_W^2 + \sigma_E^2 = V(W + E_2) = V(X_2)$ and
$Cov(X_1, X_2) = Cov(W + E_1, W + E_2) = Cov(W, W) + Cov(W, E_2) +$
$Cov(E_1, W) + Cov(E_1, E_2) = Cov(W, W) = V(W) = \sigma_w^2.$

Thus, $\rho = \dfrac{\sigma_W^2}{\sqrt{\sigma_W^2 + \sigma_E^2} \cdot \sqrt{\sigma_W^2 + \sigma_E^2}} = \dfrac{\sigma_W^2}{\sigma_W^2 + \sigma_E^2}$

b. $\rho = \dfrac{1}{1 + .0001} = .9999$

93. $E(Y) \doteq h(\mu_1, \mu_2, \mu_3, \mu_4) = 120\left[\frac{1}{10} + \frac{1}{15} + \frac{1}{20}\right] = 26$

The partial derivatives of $h(\mu_1, \mu_2, \mu_3, \mu_4)$ with respect to x_1, x_2, x_3, and x_4 are

$-\dfrac{x_4}{x_1^2}$, $-\dfrac{x_4}{x_2^2}$, $-\dfrac{x_4}{x_3^2}$, and $\dfrac{1}{x_1} + \dfrac{1}{x_2} + \dfrac{1}{x_3}$, respectively. Substituting $x_1 = 10$, $x_2 = $
15, $x_3 = 20$, and $x_4 = 120$ gives -1.2, $-.5333$, $-.3000$, and $.2167$, respectively, so V(Y)
$= (1)(-1.2)^2 + (1)(-.5333)^2 + (1.5)(-.3000)^2 + (4.0)(.2167)^2 = 2.6783$, and the
approximate sd of y is 1.64.

Chapter 5: Joint Probability Distributions and Random Samples

CHAPTER 6

Section 6.1

1.

    a. We use the sample mean, $\bar{x}$ to estimate the population mean μ.

$$\hat{\mu} = \bar{x} = \frac{\Sigma x_i}{n} = \frac{219.80}{27} = 8.1407$$

    b. We use the sample median, $\tilde{x} = 7.7$ (the middle observation when arranged in ascending order).

    c. We use the sample standard deviation,

$$s = \sqrt{s^2} = \sqrt{\frac{1860.94 - \frac{(219.8)^2}{27}}{26}} = 1.660$$

    d. With "success" = observation greater than 10, x = # of successes = 4, and
$\hat{p} = \frac{x}{n} = \frac{4}{27} = .1481$

    e. We use the sample (std dev)/(mean), or $\dfrac{s}{\bar{x}} = \dfrac{1.660}{8.1407} = .2039$

3.

    a. We use the sample mean, $\bar{x} = 1.3481$

    b. Because we assume normality, the mean = median, so we also use the sample
        mean $\bar{x} = 1.3481$. We could also easily use the sample median.

    c. We use the 90th percentile of the sample:
$$\hat{\mu} + (1.28)\hat{\sigma} = \bar{x} + 1.28s = 1.3481 + (1.28)(.3385) = 1.7814 .$$

    d. Since we can assume normality,
$$P(X < 1.5) \approx P\left(Z < \frac{1.5 - \bar{x}}{s}\right) = P\left(Z < \frac{1.5 - 1.3481}{.3385}\right) = P(Z < .45) = .6736$$

    e. The estimated standard error of $\bar{x} = \dfrac{\hat{\sigma}}{\sqrt{n}} = \dfrac{s}{\sqrt{n}} = \dfrac{.3385}{\sqrt{16}} = .0846$

5. $N = 5{,}000$ $T = 1{,}761{,}300$

        $\bar{y} = 374.6$ $\bar{x} = 340.6$ $\bar{d} = 34.0$

        $\hat{\theta}_1 = N\bar{x} = (5{,}000)(340.6) = 1{,}703{,}000$

        $\hat{\theta}_2 = T - N\bar{d} = 1{,}761{,}300 - (5{,}000)(34.0) = 1{,}591{,}300$

        $\hat{\theta}_3 = T\left(\dfrac{\bar{x}}{\bar{y}}\right) = 1{,}761{,}300\left(\dfrac{340.6}{374.6}\right) = 1{,}601{,}438.281$

Chapter 6: Point Estimation

7.

a. $\hat{\mu} = \bar{x} = \dfrac{\sum x_i}{n} = \dfrac{1206}{10} = 120.6$

b. $\hat{\tau} = 10,000 \qquad \hat{\mu} = 1,206,000$

c. 8 of 10 houses in the sample used at least 100 therms (the "successes"), so
$\hat{p} = \frac{8}{10} = .80$.

d. The ordered sample values are 89, 99, 103, 109, 118, 122, 125, 138, 147, 156, from which the two middle values are 118 and 122, so
$$\hat{\tilde{\mu}} = \tilde{x} = \dfrac{118 + 122}{2} = 120.0$$

9.

a. $E(\bar{X}) = \mu = E(X) = \lambda$, so $\bar{X}$ is an unbiased estimator for the Poisson parameter λ; $\sum x_i = (0)(18) + (1)(37) + ... + (7)(1) = 317$, since n = 150,
$$\hat{\lambda} = \bar{x} = \dfrac{317}{150} = 2.11.$$

b. $\sigma_{\bar{x}} = \dfrac{\sigma}{\sqrt{n}} = \dfrac{\sqrt{\lambda}}{\sqrt{n}}$, so the estimated standard error is $\sqrt{\dfrac{\hat{\lambda}}{n}} = \dfrac{\sqrt{2.11}}{\sqrt{150}} = .119$

11.

a. $E\left(\dfrac{X_1}{n_1} - \dfrac{X_2}{n_2}\right) = \dfrac{1}{n_1}E(X_1) - \dfrac{1}{n_2}E(X_2) = \dfrac{1}{n_1}(n_1 p_1) - \dfrac{1}{n_2}(n_2 p_2) = p_1 - p_2$

b. $Var\left(\dfrac{X_1}{n_1} - \dfrac{X_2}{n_2}\right) = Var\left(\dfrac{X_1}{n_1}\right) + Var\left(\dfrac{X_2}{n_2}\right) = \left(\dfrac{1}{n_1}\right)^2 Var(X_1) + \left(\dfrac{1}{n_2}\right)^2 Var(X_2)$

$\dfrac{1}{n_1^2}(n_1 p_1 q_1) + \dfrac{1}{n_2^2}(n_2 p_2 q_2) = \dfrac{p_1 q_1}{n_1} + \dfrac{p_2 q_2}{n_2}$, and the standard error is the square root of this quantity.

c. With $\hat{p}_1 = \dfrac{x_1}{n_1}$, $\hat{q}_1 = 1 - \hat{p}_1$, $\hat{p}_2 = \dfrac{x_2}{n_2}$, $\hat{q}_2 = 1 - \hat{p}_2$, the estimated standard

error is $\sqrt{\dfrac{\hat{p}_1\hat{q}_1}{n_1} + \dfrac{\hat{p}_2\hat{q}_2}{n_2}}$.

d. $(\hat{p}_1 - \hat{p}_2) = \dfrac{127}{200} - \dfrac{176}{200} = .635 - .880 = -.245$

e. $\sqrt{\dfrac{(.635)(.365)}{200} + \dfrac{(.880)(.120)}{200}} = .041$

13. $E(X) = \int_{-1}^{1} x \cdot \frac{1}{2}(1 + \theta x)dx = \dfrac{x^2}{4} + \dfrac{\theta x^3}{6}\Big|_{-1}^{1} = \dfrac{1}{3}\theta \qquad E(X) = \dfrac{1}{3}\theta$

$E(\overline{X}) = \dfrac{1}{3}\theta \quad \hat{\theta} = 3\overline{X} \Rightarrow E(\hat{\theta}) = E(3\overline{X}) = 3E(\overline{X}) = 3\left(\dfrac{1}{3}\right)\theta = \theta$

15.

a. $E(X^2) = 2\theta$ implies that $E\left(\dfrac{X^2}{2}\right) = \theta$. Consider $\hat{\theta} = \dfrac{\sum X_i^2}{2n}$. Then

$E(\hat{\theta}) = E\left(\dfrac{\sum X_i^2}{2n}\right) = \dfrac{\sum E(X_i^2)}{2n} = \dfrac{\sum 2\theta}{2n} = \dfrac{2n\theta}{2n} = \theta$, implying that $\hat{\theta}$ is

an unbiased estimator for θ.

b. $\sum x_i^2 = 1490.1058$, so $\hat{\theta} = \dfrac{1490.1058}{20} = 74.505$

17.

a. $E(\hat{p}) = \sum_{x=0}^{\infty} \frac{r-1}{x+r-1} \cdot \binom{x+r-1}{x} \cdot p^r \cdot (1-p)^x$

$= p \sum_{x=0}^{\infty} \frac{(x+r-2)!}{x!(r-2)!} \cdot p^{r-1} \cdot (1-p)^x = p \sum_{x=0}^{\infty} \binom{x+r-2}{x} p^{r-1} (1-p)^x$

$= p \sum_{x=0}^{\infty} nb(x;r-1,p) = p$.

b. For the given sequence, x = 5, so $\hat{p} = \dfrac{5-1}{5+5-1} = \dfrac{4}{9} = .444$

19.

a. $\lambda = .5p + .15 \Rightarrow 2\lambda = p + .3$, so $p = 2\lambda - .3$ and

$\hat{p} = 2\hat{\lambda} - .3 = 2\left(\dfrac{Y}{n}\right) - .3$; the estimate is $2\left(\dfrac{20}{80}\right) - .3 = .2$.

b. $E(\hat{p}) = E\left(2\hat{\lambda} - .3\right) = 2E\left(\hat{\lambda}\right) - .3 = 2\lambda - .3 = p$, as desired.

c. Here $\lambda = .7p + (.3)(.3)$, so $p = \dfrac{10}{7}\lambda - \dfrac{9}{70}$ and $\hat{p} = \dfrac{10}{7}\left(\dfrac{Y}{n}\right) - \dfrac{9}{70}$.

Section 6.2

21.

a. $E(X) = \beta \cdot \Gamma\left(1 + \dfrac{1}{\alpha}\right)$ and

$E(X^2) = Var(X) + [E(X)]^2 = \beta^2 \Gamma\left(1 + \dfrac{2}{\alpha}\right)$, so the moment estimators

$\hat{\alpha}$ and $\hat{\beta}$ are the solution to $\overline{X} = \hat{\beta} \cdot \Gamma\left(1 + \dfrac{1}{\hat{\alpha}}\right)$,

$\dfrac{1}{n}\sum X_i^2 = \hat{\beta}^2 \Gamma\left(1 + \dfrac{2}{\hat{\alpha}}\right)$. Thus $\hat{\beta} = \dfrac{\overline{X}}{\Gamma\left(1 + \dfrac{1}{\hat{\alpha}}\right)}$, so once $\hat{\alpha}$ has been

determined $\Gamma\left(1 + \dfrac{1}{\hat{\alpha}}\right)$ is evaluated and $\hat{\beta}$ then computed. Since

$\overline{X}^2 = \hat{\beta}^2 \cdot \Gamma^2\left(1 + \dfrac{1}{\hat{\alpha}}\right)$, $\dfrac{1}{n}\sum \dfrac{X_i^2}{\overline{X}^2} = \dfrac{\Gamma\left(1 + \dfrac{2}{\hat{\alpha}}\right)}{\Gamma^2\left(1 + \dfrac{1}{\hat{\alpha}}\right)}$, so this equation must be

solved to obtain $\hat{\alpha}$.

b. From **a**, $\dfrac{1}{20}\left(\dfrac{16,500}{28.0^2}\right) = 1.05 = \dfrac{\Gamma\left(1 + \dfrac{2}{\hat{\alpha}}\right)}{\Gamma^2\left(1 + \dfrac{1}{\hat{\alpha}}\right)}$, so

$\dfrac{1}{1.05} = .95 = \dfrac{\Gamma^2\left(1 + \dfrac{1}{\hat{\alpha}}\right)}{\Gamma\left(1 + \dfrac{2}{\hat{\alpha}}\right)}$, and from the hint, $\dfrac{1}{\hat{\alpha}} = .2 \Rightarrow \hat{\alpha} = 5$. Then

$\hat{\beta} = \dfrac{\overline{x}}{\Gamma(1.2)} = \dfrac{28.0}{\Gamma(1.2)}$.

23. For a single sample from a Poisson distribution,

$$f(x_1,...,x_n;\lambda) = \frac{e^{-\lambda}\lambda^{x_1}}{x_1!}\cdots\frac{e^{-\lambda}\lambda^{x_n}}{x_n!} = \frac{e^{-n\lambda}\lambda^{\sum x_i}}{x_1!...x_n!}, \text{ so}$$

$$\ln[f(x_1,...,x_n;\lambda)] = -n\lambda + \sum x_i \ln(\lambda) - \sum \ln(x_i!). \text{ Thus}$$

$$\frac{d}{d\lambda}[\ln[f(x_1,...,x_n;\lambda)]] = -n + \frac{\sum x_i}{\lambda} = 0 \Rightarrow \hat{\lambda} = \frac{\sum x_i}{n} = \bar{x}. \text{ For our}$$

problem, $f(x_1,...,x_n,y_1...y_n;\lambda_1,\lambda_2)$ is a product of the x sample likelihood and

the y sample likelihood, implying that $\hat{\lambda}_1 = \bar{x}, \hat{\lambda}_2 = \bar{y}$, and (by the invariance

principle) $(\hat{\lambda_1 - \lambda_2}) = \bar{x} - \bar{y}$.

25.

    a. $\hat{\mu} = \bar{x} = 384.4; s^2 = 395.16$, so

$$\frac{1}{n}\sum(x_i - \bar{x})^2 = \hat{\sigma}^2 = \frac{9}{10}(395.16) = 355.64 \text{ and}$$

$$\hat{\sigma} = \sqrt{355.64} = 18.86 \text{ (this is not s).}$$

    b. The 95th percentile is $\mu + 1.645\sigma$, so the mle of this is (by the invariance
    principle) $\hat{\mu} + 1.645\hat{\sigma} = 415.42$.

27.

a. $f(x_1,...,x_n;\alpha,\beta) = \dfrac{(x_1 x_2 ... x_n)^{\alpha-1} e^{-\Sigma x_i / \beta}}{\beta^{n\alpha} \Gamma^n(\alpha)}$, so the log likelihood is

$(\alpha-1)\sum \ln(x_i) - \dfrac{\sum x_i}{\beta} - n\alpha \ln(\beta) - n \ln \Gamma(\alpha)$. Equating both $\dfrac{d}{d\alpha}$ and

$\dfrac{d}{d\beta}$ to 0 yields $\sum \ln(x_i) - n\ln(\beta) - n\dfrac{d}{d\alpha}\Gamma(\alpha) = 0$ and

$\dfrac{\sum x_i}{\beta^2} = \dfrac{n\alpha}{\beta} = 0$, a very difficult system of equations to solve.

b. From the second equation in **a**, $\dfrac{\sum x_i}{\beta} = n\alpha \Rightarrow \bar{x} = \alpha\beta = \mu$, so the mle of

μ is $\hat{\mu} = \bar{X}$.

29.

a. The joint pdf (likelihood function) is

$$f(x_1,...,x_n;\lambda,\theta) = \begin{cases} \lambda^n e^{-\lambda\Sigma(x_i-\theta)} & x_1 \ge \theta,...,x_n \ge \theta \\ 0 & otherwise \end{cases}$$

Notice that $x_1 \ge \theta,...,x_n \ge \theta$ iff $\min(x_i) \ge \theta$,

and that $-\lambda\Sigma(x_i - \theta) = -\lambda\Sigma x_i + n\lambda\theta$.

Thus likelihood $= \begin{cases} \lambda^n \exp(-\lambda\Sigma x_i) \exp(n\lambda\theta) & \min(x_i) \ge \theta \\ 0 & \min(x_i) < \theta \end{cases}$

Consider maximization wrt θ. Because the exponent $n\lambda\theta$ is positive,
increasing θ will increase the likelihood provided that $\min(x_i) \ge \theta$; if we
make θ larger than $\min(x_i)$, the likelihood drops to 0. This implies that the
mle of θ is $\hat{\theta} = \min(x_i)$. The log likelihood is now $n\ln(\lambda) - \lambda\Sigma(x_i - \hat{\theta})$.
Equating the derivative wrt λ to 0 and solving yields

$$\hat{\lambda} = \dfrac{n}{\Sigma(x_i - \hat{\theta})} = \dfrac{n}{\Sigma x_i - n\hat{\theta}}.$$

b. $\hat{\theta} = \min(x_i) = .64$, and $\Sigma x_i = 55.80$, so $\hat{\lambda} = \dfrac{10}{55.80 - 6.4} = .202$

Supplementary Exercises

31. $P(|\overline{X} - \mu| > \varepsilon) = P(\overline{X} - \mu > \varepsilon) + P(\overline{X} - \mu < -\varepsilon) = P\left(\dfrac{\overline{X} - \mu}{\sigma/\sqrt{n}} > \dfrac{\varepsilon}{\sigma/\sqrt{n}}\right) + P\left(\dfrac{\overline{X} - \mu}{\sigma/\sqrt{n}} < \dfrac{-\varepsilon}{\sigma/\sqrt{n}}\right)$

$= P\left(Z > \dfrac{\sqrt{n}\varepsilon}{\sigma}\right) + P\left(Z < \dfrac{-\sqrt{n}\varepsilon}{\sigma}\right) = \int_{\sqrt{n}\varepsilon/\sigma}^{\infty} \dfrac{1}{\sqrt{2\pi}} e^{-z^2/2} dz + \int_{-\infty}^{-\sqrt{n}\varepsilon/\sigma} \dfrac{1}{\sqrt{2\pi}} e^{-z^2/2} dz$

As $n \to \infty$, both integrals $\to 0$ since $\displaystyle\lim_{c \to \infty} \int_c^{\infty} \dfrac{1}{\sqrt{2\pi}} e^{-z^2/2} dz = 0$.

33. Let x_1 = the time until the first birth, x_2 = the elapsed time between the first and second births, and so on. Then

$$f(x_1,...,x_n;\lambda) = \lambda e^{-\lambda x_1} \cdot (2\lambda)e^{-2\lambda x_2} ...(n\lambda)e^{-n\lambda x_n} = n!\,\lambda^n e^{-\lambda \Sigma k x_k}.$$ Thus the log

likelihood is $\ln(n!) + n\ln(\lambda) - \lambda\Sigma k x_k$. Taking $\dfrac{d}{d\lambda}$ and equating to 0 yields

$\hat{\lambda} = \dfrac{n}{\displaystyle\sum_{k=1}^{n} k x_k}$. For the given sample, n = 6, x_1 = 25.2, x_2 = 41.7 – 25.2 = 16.5, x_3 =

9.5, x_4 = 4.3, x_5 = 4.0, x_6 = 2.3; so

$$\sum_{k=1}^{6} k x_k = (1)(25.2) + (2)(16.5) + ... + (6)(2.3) = 137.7 \text{ and}$$

$\hat{\lambda} = \dfrac{6}{137.7} = .0436$.

35.

$x_i + x_j$	23.5	26.3	28.0	28.2	29.4	29.5	30.6	31.6	33.9	49.3
23.5	23.5	24.9	25.75	25.85	26.45	26.5	27.05	27.55	28.7	36.4
26.3		26.3	27.15	27.25	27.85	27.9	28.45	28.95	30.1	37.8
28.0			28.0	28.1	28.7	28.75	29.3	29.8	30.95	38.65
28.2				28.2	28.8	28.85	29.4	29.9	31.05	38.75
29.4					29.4	29.45	30.0	30.5	30.65	39.35
29.5						29.5	30.05	30.55	31.7	39.4
30.6							30.6	31.1	32.25	39.95
31.6								31.6	32.75	40.45
33.9									33.9	41.6
49.3										49.3

There are 55 averages, so the median is the 28th in order of increasing magnitude. Therefore, $\hat{\mu} = 29.5$

37. Let $c = \dfrac{\Gamma\left(\frac{n-1}{2}\right)}{\Gamma\left(\frac{n}{2}\right)\cdot\sqrt{\frac{2}{n-1}}}$. Then E(cS) = cE(S), and c cancels with the two Γ factors and

the square root in E(S), leaving just σ. When n = 20, $c = \dfrac{\Gamma(9.5)}{\Gamma(10)\cdot\sqrt{\frac{2}{19}}}$.

$\Gamma(10) = 9!$ and $\Gamma(9.5) = (8.5)(7.5)...(1.5)(.5)\Gamma(.5)$, but $\Gamma(.5) = \sqrt{\pi}$.
Straightforward calculation gives c = 1.0132.

CHAPTER 7

Section 7.1

1.

    a. $z_{\alpha/2} = 2.81$ implies that $\alpha/2 = 1 - \Phi(2.81) = .0025$, so $\alpha = .005$ and the confidence level is $100(1 - \alpha)\% = 99.5\%$.

    b. $z_{\alpha/2} = 1.44$ for $\alpha = 2[1 - \Phi(1.44)] = .15$, and $100(1 - \alpha)\% = 85\%$.

    c.   99.7% implies that $\alpha = .003$, $\alpha/2 = .0015$, and $z_{.0015} = 2.96$. (Look for cumulative area .9985 in the main body of table A.3, the Z table.)

    d.   75% implies $\alpha = .25$, $\alpha/2 = .125$, and $z_{.125} = 1.15$.

3.

    a.   A 90% confidence interval will be narrower (See 2b, above) Also, the z critical value for a 90% confidence level is 1.645, smaller than the z of 1.96 for the 95% confidence level, thus producing a narrower interval.

    b.   Not a correct statement. Once and interval has been created from a sample, the mean μ is either enclosed by it, or not. The 95% confidence is in the general procedure, for repeated sampling.

    c.   Not a correct statement. The interval is an estimate for the population mean, not a boundary for population values.

    d.   Not a correct statement. In theory, if the process were repeated an infinite number of times, 95% of the intervals would contain the population mean μ.

5.

a. $4.85 \pm \dfrac{(1.96)(.75)}{\sqrt{20}} = 4.85 \pm .33 = (4.52, 5.18)$.

b. $z_{\alpha/2} = z_{.02/2} = z_{.01} = 2.33$, so the interval is $4.56 \pm \dfrac{(2.33)(.75)}{\sqrt{16}} = (4.12, 5.00)$.

c. $n = \left[\dfrac{2(1.96)(.75)}{.40} \right]^2 = 54.02$, so n = 55.

d. $n = \left[\dfrac{2(2.58)(.75)}{.2} \right]^2 = 93.61$, so n = 94.

7. If $L = 2z_{\alpha/2} \dfrac{\sigma}{\sqrt{n}}$ and we increase the sample size by a factor of 4, the new length is

$L' = 2z_{\alpha/2} \dfrac{\sigma}{\sqrt{4n}} = \left[2z_{\alpha/2} \dfrac{\sigma}{\sqrt{n}} \right]\left(\dfrac{1}{2} \right) = \dfrac{L}{2}$. Thus halving the length requires n to be

increased fourfold. If $n' = 25n$, then $L' = \dfrac{L}{5}$, so the length is decreased by a

factor of 5.

9.

a. $\left(\bar{x} - 1.645\dfrac{\sigma}{\sqrt{n}}, \infty\right)$. From 5a, $\bar{x} = 4.85$, $\sigma = .75$, $n = 20$;

$4.85 - 1.645\dfrac{.75}{\sqrt{20}} = 4.5741$, so the interval is $\left(4.5741, \infty\right)$.

b. $\left(\bar{x} - z_\alpha\dfrac{\sigma}{\sqrt{n}}, \infty\right)$

c. $\left(-\infty, \bar{x} + z_\alpha\dfrac{\sigma}{\sqrt{n}}\right)$; From 4a, $\bar{x} = 58.3$, $\sigma = 3.0$, $n = 25$;

$58.3 + 2.33\dfrac{3}{\sqrt{25}} = \left(-\infty, 59.70\right)$

11. Y is a binomial r.v. with n = 1000 and p = .95, so E(Y) = np = 950, the expected number of intervals that capture μ, and $\sigma_Y = \sqrt{npq} = 6.892$. Using the normal approximation to the binomial distribution, P(940 ≤ Y ≤ 960) = P(939.5 ≤ Y_{normal} ≤ 960.5) = P(-1.52 ≤ Z ≤ 1.52) = .9357 - .0643 = .8714.

Section 7.2

13.

a. $\bar{x} \pm z_{.025}\dfrac{s}{\sqrt{n}} = 1.028 \pm 1.96\dfrac{.163}{\sqrt{69}} = 1.028 \pm .038 = \left(.990, 1.066\right)$

b. $w = .05 = \dfrac{2(1.96)(.16)}{\sqrt{n}} \Rightarrow \sqrt{n} = \dfrac{2(1.96)(.16)}{.05} = 12.544$

$\Rightarrow n = (12.544)^2 \approx 158$

15.

    a. $z_\alpha = .84$, and $\Phi(.84) = .7995 \approx .80$, so the confidence level is 80%.

    b. $z_\alpha = 2.05$, and $\Phi(2.05) = .9798 \approx .98$, so the confidence level is 98%.

    c. $z_\alpha = .67$, and $\Phi(.67) = .7486 \approx .75$, so the confidence level is 75%.

17. $\bar{x} - z_{.01} \dfrac{s}{\sqrt{n}} = 135.39 - 2.33 \dfrac{4.59}{\sqrt{153}} = 135.39 - .865 = 134.53$ With a

confidence level of 99%, the true average ultimate tensile strength is between $(134.53, \infty)$.

19. $\hat{p} = \dfrac{201}{356} = .5646$; We calculate a 95% confidence interval for the proportion of all

dies that pass the probe:

$$\frac{.5646 + \dfrac{(1.96)^2}{2(356)} \pm 1.96 \sqrt{\dfrac{(.5646)(.4354)}{356} + \dfrac{(1.96)^2}{4(356)^2}}}{1 + \dfrac{(1.96)^2}{356}} = \frac{.5700 \pm .0518}{1.01079} = (.513, .615)$$

21. $\hat{p} = \dfrac{133}{539} = .2468$; the 95% lower confidence bound is:

$$\frac{.2468 + \dfrac{(1.645)^2}{2(539)} - 1.645 \sqrt{\dfrac{(.2468)(.7532)}{539} + \dfrac{(1.645)^2}{4(539)^2}}}{1 + \dfrac{(1.645)^2}{539}} = \frac{.2493 - .0307}{1.005} = .218$$

23.

a. $\hat{p} = \dfrac{24}{37} = .6486$; The 99% confidence interval for p is

$$\dfrac{.6486 + \dfrac{(2.58)^2}{2(37)} \pm 2.58 \sqrt{\dfrac{(.6486)(.3514)}{37} + \dfrac{(2.58)^2}{4(37)^2}}}{1 + \dfrac{(2.58)^2}{37}} = \dfrac{.7386 \pm .2216}{1.1799} = (.438, .814)$$

b. $n = \dfrac{2(2.58)^2(.25) - (2.58)^2(.01) \pm \sqrt{4(2.58)^4(.25)(.25 - .01) + .01(2.58)^4}}{.01}$

$$= \dfrac{3.261636 \pm 3.3282}{.01} \approx 659$$

25.

a. $n = \dfrac{2(1.96)^2(.25) - (1.96)^2(.01) \pm \sqrt{4(1.96)^4(.25)(.25 - .01) + .01(1.96)^4}}{.01} \approx 381$

b. $n = \dfrac{2(1.96)^2\left(\frac{1}{3} \cdot \frac{2}{3}\right) - (1.96)^2(.01) \pm \sqrt{4(1.96)^4\left(\frac{1}{3} \cdot \frac{2}{3}\right)\left(\frac{1}{3} \cdot \frac{2}{3} - .01\right) + .01(1.96)^4}}{.01} \approx 339$

27.   Note that the midpoint of the new interval is $\dfrac{x+\frac{z^2}{2}}{n+z^2}$, which is roughly $\dfrac{x+2}{n+4}$ with

a confidence level of 95% and approximating $1.96 \approx 2$. The variance of this

quantity is $\dfrac{np(1-p)}{\left(n+z^2\right)^2}$, or roughly $\dfrac{p(1-p)}{n+4}$. Now replacing p with $\dfrac{x+2}{n+4}$, we

have $\left(\dfrac{x+2}{n+4}\right) \pm z_{\alpha/2} \sqrt{\dfrac{\left(\dfrac{x+2}{n+4}\right)\left(1-\dfrac{x+2}{n+4}\right)}{n+4}}$; For clarity, let $x^* = x+2$ and

$n^* = n+4$, then $\hat{p}^* = \dfrac{x^*}{n^*}$ and the formula reduces to $\hat{p}^* \pm z_{\alpha/2} \sqrt{\dfrac{\hat{p}^* \hat{q}^*}{n^*}}$, the

desired conclusion. For further discussion, see the Agresti article.

Section 7.3

29.

a. $t_{.025,10} = 2.228$

d. $t_{.005,50} = 2.678$

b. $t_{.025,20} = 2.086$

e. $t_{.01,25} = 2.485$

c. $t_{.005,20} = 2.845$

f. $-t_{.025,5} = -2.571$

31.

a. $t_{.05,10} = 1.812$

d. $t_{.01,4} = 3.747$

b. $t_{.05,15} = 1.753$

e. $\approx t_{.025,24} = 2.064$

c. $t_{.01,15} = 2.602$

f. $t_{.01,37} \approx 2.429$

Chapter 7: Statistical Intervals Based on a Single Sample

33.

    a. The boxplot indicates a very slight positive skew, with no outliers. The data appears to center near 438.

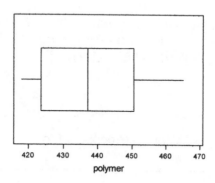

    polymer

    b. Based on a normal probability plot, it is reasonable to assume the sample observations came from a normal distribution.

    c. With d.f. $= n - 1 = 16$, the critical value for a 95% C.I. is $t_{.025,16} = 2.120$, and the interval is

$$438.29 \pm (2.120)\left(\frac{15.14}{\sqrt{17}}\right) = 438.29 \pm 7.785 = (430.51, 446.08).$$ Since

440 is within the interval, 440 is a plausible value for the true mean. 450, however, is not, since it lies outside the interval.

35. $n = 5$, $\bar{x} = 2887.6$, $s = .84.0$; $t_{.025,4} = 2.776$

    a. A 95% C.I. for the mean: $2887.6 \pm (2.776)\left(\dfrac{84}{\sqrt{5}}\right) \Rightarrow (2783.3, 2991.9)$

    b. A 95% Prediction Interval:

$$2887.6 \pm 2.776(84)\sqrt{1 + \frac{1}{5}} \Rightarrow (2632.1, 3143.1).$$ The P.I. is considerably

larger than the C.I., about 2.5 times larger.

37.

    a. A 95% C.I. : $.9255 \pm 2.093(.0181) = .9255 \pm .0379 \Rightarrow (.8876, .9634)$

    b. A 95% P.I. :

        $.9255 \pm 2.093(.0809)\sqrt{1 + \frac{1}{20}} = .9255 \pm .1735 \Rightarrow (.7520, 1.0990)$

    c. A tolerance interval is requested, with k = 99, confidence level 95%, and n = 20. The tolerance critical value, from Table A.6, is 3.615. The interval is

        $.9255 \pm 3.615(.0809) \Rightarrow (.6330, 1.2180)$.

39.

    a.

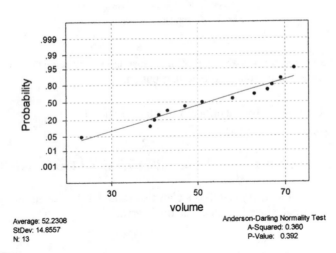

Average: 52.2308
StDev: 14.8557
N: 13

Anderson-Darling Normality Test
A-Squared: 0.360
P-Value: 0.392

Based on the above plot, generated by Minitab, it is plausible that the population distribution is normal.

    b. We require a tolerance interval. (from table A6, with 95% confidence, k = 95, and n=13, the tcv = 3.081.

        $\bar{x} \pm (tcv)s = 52.231 \pm 3.081(14.856) = 52.231 \pm 45.771$

        $\Rightarrow (6.460, 98.002)$

c. A prediction interval, with $t_{.025,12} = 2.179$:

$$52.231 \pm 2.179(14.856)\sqrt{1 + \tfrac{1}{13}} = 52.231 \pm 33.593 \Rightarrow (18.638, 85.824)$$

41. The 20 d.f. row of Table A.5 shows that 1.725 captures upper tail area .05 and 1.325 captures uppertail area .10 The confidence level for each interval is 100(central area)%. For the first interval, central area = 1 – sum of tail areas = 1 – (.25 + .05) = .70, and for the second and third intervals the central areas are 1 – (.20 + .10) = .70 and 1 – (.15 + .15) = 70. Thus each interval has confidence level 70%. The width of the first interval is $\dfrac{s(.687 + 1.725)}{\sqrt{n}} = \dfrac{.2412s}{\sqrt{n}}$, whereas the widths of the second and third intervals are 2.185 and 2.128 respectively. The third interval, with symmetrically placed critical values, is the shortest, so it should be used. This will always be true for a t interval.

Section 7.2

43.

a. $\chi^2_{.05,10} = 18.307$

b. $\chi^2_{.95,10} = 3.940$

c. Since $10.987 = \chi^2_{.975,22}$ and
$$36.78 = \chi^2_{.025,22}, P\left(\chi^2_{.975,22} \le \chi^2 \le \chi^2_{.025,22}\right) = .95.$$

d. Since $14.61 = \chi^2_{.95,25}$ and $37.65 = \chi^2_{.05,25}, P\left(\chi^2_{.95,25} \le \chi^2 \le \chi^2_{.05,25}\right) = .90.$

45. n = 22 implies that d.f. = n – 1 = 21, so the .995 and .005 columns of Table A.7 give the necessary chi-squared critical values as 8.033 and 41.399. $\Sigma x_i = 1701.3$ and $\Sigma x_i^2 = 132,097.35$, so $s^2 = 25.368$. The interval for σ^2 is
$$\left(\frac{21(25.368)}{41.399}, \frac{21(25.368)}{8.033}\right) = (12.868, 66.317) \text{ and that for } \sigma \text{ is } (3.6, 8.1)$$
Validity of this interval requires that fracture toughness be (at least approximately) normally distributed.

Supplementary Exercises

47.

a. $n = 48$, $\bar{x} = 8.079$, $s^2 = 23.7017$, and $s = 4.868$.
A 95% C.I. for μ = the true average strength is

$$\bar{x} \pm 1.96 \frac{s}{\sqrt{n}} = 8.079 \pm 1.96 \frac{4.868}{\sqrt{48}} = 8.079 \pm 1.377 = (6.702, 9.456)$$

b. $\hat{p} = \frac{13}{48} = .2708$. A 95% C.I. is

$$\frac{.2708 + \frac{1.96^2}{2(48)} \pm 1.96\sqrt{\frac{(.2708)(.7292)}{48} + \frac{1.96^2}{4(48)^2}}}{1 + \frac{1.96^2}{48}} = \frac{.3108 \pm .1319}{1.0800} = (.166, .410)$$

49.

a. There appears to be a slight positive skew in the middle half of the sample, but the lower whisker is much longer than the upper whisker. The extent of variability is rather substantial, although there are no outliers.

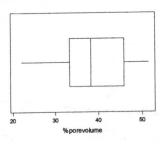

b. The pattern of points in a normal probability plot is reasonably linear, so, yes, normality is plausible.

c. $n = 18$, $\bar{x} = 38.66$, $s = 8.473$, and $t_{.01,17} = 2.586$. The 98% confidence interval is $38.66 \pm 2.586 \frac{8.473}{\sqrt{18}} = 38.66 \pm 5.13 = (33.53, 43.79)$.

51.

a. $\hat{p} = \dfrac{136}{200} = .680 \Rightarrow$ a 90% C.I. is

$$\dfrac{.680 + \dfrac{1.645^2}{2(200)} \pm 1.645 \sqrt{\dfrac{(.680)(.320)}{200} + \dfrac{1.645^2}{4(200)^2}}}{1 + \dfrac{1.645^2}{200}} = \dfrac{.6868 \pm .0547}{1.01353} = (.624, .732)$$

b. $n = \dfrac{2(1.645)^2(.25) - (1.645)^2(.05)^2 \pm \sqrt{4(1.645)^4(.25)(.25 - .0025) + .05^2(1.645)^4}}{.0025}$

$= \dfrac{1.3462 \pm 1.3530}{.0025} = 1079.7 \Rightarrow$ use n = 1080

c. No, it gives a 95% upper bound.

53. With $\hat{\theta} = \frac{1}{3}\left(\overline{X}_1 + \overline{X}_2 + \overline{X}_3\right) - \overline{X}_4$, $\sigma_{\hat{\theta}}^2 = \frac{1}{9} Var\left(\overline{X}_1 + \overline{X}_2 + \overline{X}_3\right) + Var\left(\overline{X}_4\right)$

$= \dfrac{1}{9}\left(\dfrac{\sigma_1^2}{n_1} + \dfrac{\sigma_2^2}{n_2} + \dfrac{\sigma_3^2}{n_3}\right) + \dfrac{\sigma_4^2}{n_4}$; $\hat{\sigma}_{\hat{\theta}}$ is obtained by replacing each $\hat{\sigma}_i^2$ by s_i^2 and

taking the square root. The large-sample interval for θ is then

$\frac{1}{3}\left(\overline{x}_1 + \overline{x}_2 + \overline{x}_3\right) - \overline{x}_4 \pm z_{\alpha/2}\sqrt{\dfrac{1}{9}\left(\dfrac{s_1^2}{n_1} + \dfrac{s_2^2}{n_2} + \dfrac{s_3^2}{n_3}\right) + \dfrac{s_4^2}{n_4}}$. For the given data,

$\hat{\theta} = -.50$, $\hat{\sigma}_{\hat{\theta}} = .1718$, so the interval is $-.50 \pm 1.96(.1718) = (-.84, -.16)$.

55. The specified condition is that the interval be length .2, so

$$n = \left[\dfrac{2(1.96)(.8)}{.2}\right]^2 = 245.86, \text{ so n = 246 should be used.}$$

57. Proceeding as in Example 7.5 with T_r replacing ΣX_i, the C.I. for $\dfrac{1}{\lambda}$ is

$$\left(\frac{2t_r}{\chi^2_{1-\alpha/2,2r}}, \frac{2t_r}{\chi^2_{\alpha/2,2r}} \right)$$ where $t_r = y_1 + \ldots + y_r + (n-r)y_r$. In Example 6.7, n = 20, r = 10, and t_r = 1115. With d.f. = 20, the necessary critical values are 9.591 and 34.170, giving the interval (65.3, 232.5). This is obviously an extremely wide interval. The censored experiment provides less information about $\frac{1}{\lambda}$ than would an uncensored experiment with n = 20.

59.

a. $\displaystyle\int_{(\alpha/2)^{1/n}}^{(1-\alpha/2)^{1/n}} nu^{n-1}du = u^n \Big]_{(\alpha/2)^{1/n}}^{(1-\alpha/2)^{1/n}} = 1 - \frac{\alpha}{2} - \frac{\alpha}{2} = 1 - \alpha$. From the

probability statement, $\dfrac{(\alpha/2)^{1/n}}{\max(X_i)} \le \dfrac{1}{\theta} \le \dfrac{(1-\alpha/2)^{1/n}}{\max(X_i)}$ with probability $1 - \alpha$, so

taking the reciprocal of each endpoint and interchanging gives the C.I.

$$\left(\frac{\max(X_i)}{(1-\alpha/2)^{1/n}}, \frac{\max(X_i)}{(\alpha/2)^{1/n}} \right) \text{ for } \theta.$$

b. $\alpha^{1/n} \le \dfrac{\max(X_i)}{\theta} \le 1$ with probability $1 - \alpha$, so $1 \le \dfrac{\theta}{\max(X_i)} \le \dfrac{1}{\alpha^{1/n}}$ with

probability $1 - \alpha$, which yields the interval $\left(\max(X_i), \dfrac{\max(X_i)}{\alpha^{1/n}} \right)$.

c. It is easily verified that the interval of **b** is shorter – draw a graph of $f_U(u)$ and verify that the shortest interval which captures area $1 - \alpha$ under the curve is the rightmost such interval, which leads to the C.I. of **b**. With $\alpha = .05$, n = 5, max(x_I)=4.2; this yields (4.2, 7.65).

61. $\tilde{x} = 76.2$, the lower and upper fourths are 73.5 and 79.7, respectively, and $f_s = 6.2$. The robust interval is

$$76.2 \pm (1.93)\left(\frac{6.2}{\sqrt{22}}\right) = 76.2 \pm 2.6 = (73.6, 78.8).$$

$\bar{x} = 77.33$, s = 5.037, and $t_{.025,21} = 2.080$, so the t interval is

$$77.33 \pm (2.080)\left(\frac{5.037}{\sqrt{22}}\right) = 77.33 \pm 2.23 = (75.1, 79.6). \text{ The t interval is}$$

centered at $\bar{x}$, which is pulled out to the right of $\tilde{x}$ by the single mild outlier 93.7; the interval widths are comparable.

Chapter 7: Statistical Intervals Based on a Single Sample

CHAPTER 8

Section 8.1

1.

    a. Yes. It is an assertion about the value of a parameter.

    b. No. The sample median $\tilde{X}$ is not a parameter.

    c. No. The sample standard deviation s is not a parameter.

    d. Yes. The assertion is that the standard deviation of population #2 exceeds that of population #1

    e. No. $\overline{X}$ and $\overline{Y}$ are statistics rather than parameters, so cannot appear in a hypothesis.

    f. Yes. H is an assertion about the value of a parameter.

3.   In this formulation, H_o states the welds do not conform to specification. This assertion will not be rejected unless there is strong evidence to the contrary. Thus the burden of proof is on those who wish to assert that the specification is satisfied. Using H_a: $\mu < 100$ results in the welds being believed in conformance unless provided otherwise, so the burden of proof is on the non-conformance claim.

5.   Let σ denote the population standard deviation. The appropriate hypotheses are $H_o : \sigma = .05$ vs $H_a : \sigma < .05$. With this formulation, the burden of proof is on the data to show that the requirement has been met (the sheaths will not be used unless H_o can be rejected in favor of H_a. Type I error: Conclude that the standard deviation is $< .05$ mm when it is really equal to .05 mm. Type II error: Conclude that the standard deviation is .05 mm when it is really $< .05$.

7.   A type I error here involves saying that the plant is not in compliance when in fact it is. A type II error occurs when we conclude that the plant is in compliance when in fact it isn't. Reasonable people may disagree as to which of the two errors is more serious. If in your judgement it is the type II error, then the reformulation $H_o : \mu = 150$ vs $H_a : \mu < 150$ makes the type I error more serious.

9.

    a. R_1 is most appropriate, because x either too large or too small contradicts p = .5 and supports p ≠ .5.

    b.   A type I error consists of judging one of the tow candidates favored over the other when in fact there is a 50-50 split in the population. A type II error involves judging the split to be 50-50 when it is not.

    c.   X has a binomial distribution with n = 25 and p = 0.5. α = P(type I error) = $P(X \le 7 or X \ge 18$ when $X \sim$ Bin(25, .5)) = B(7; 25,.5) + 1 – B(17; 25,.5) = .044

    d. $\beta(.4) = P(8 \le X \le 17$ when p = .4) = B(17; 25,.5) – B(7, 25,.4) = 0.845, and $\beta(.6) = 0.845$ also. $\beta(.3) = B(17;25,.3) – B(7;25,.3) = .488 = \beta(.7)$

    e.   x = 6 is in the rejection region R_1 , so H_o is rejected in favor of H_a.

11.

    a. $H_o : \mu = 10$ vs $H_a : \mu \ne 10$

    b. α = P(rejecting H_o when H_o is true) = $P(\bar{x} \ge 10.1032$ or $\le 9.8968 when \mu = 10)$. Since $\bar{x}$ is normally distributed with standard deviation

$$\frac{\sigma}{\sqrt{n}} = \frac{.2}{5} = .04, \alpha = P(z \ge 2.58 or \le -2.58) = .005 + .005 = .01$$

    c.   When $\mu = 10.1$, $E(\bar{x}) = 10.1$, so $\beta(10.1) = P(9.8968 < \bar{x} < 10.1032$ when $\mu = 10.1) = P(-5.08 < z < .08) = .5319$. Similarly, $\beta(9.8) = P(2.42 < z < 7.58) = .0078$

    d. $c = \pm 2.58$

    e.   Now $\frac{\sigma}{\sqrt{n}} = \frac{.2}{3.162} = .0632$. Thus 10.1032 is replaced by c, where

$$\frac{c-10}{.0632} = 1.96$$ and so c = 10.124. Similarly, 9.8968 is replaced by 9.876.

f. $\bar{x} = 10.020$. Since $\bar{x}$ is neither ≥ 10.124 nor ≤ 9.876, it is not in the rejection region. H_o is not rejected; it is still plausible that $\mu = 10$.

g. $\bar{x} \geq 10.1032$ or ≤ 9.8968 iff $z \geq 2.58$ or ≤ -2.58.

13.

a. $P(\bar{x} \geq \mu_o + 2.33\dfrac{\sigma}{\sqrt{n}}\ when\mu = \mu_o) = P\left(z \geq \dfrac{\left(\mu_o + 2.33\dfrac{\sigma}{\sqrt{n}}\right)}{\dfrac{\sigma}{\sqrt{n}}}\right)$

$= P(z \geq 2.33) = .01$, where Z is a standard normal r.v.

b. P(rejecting H_o when $\mu = 99$) = $P(\bar{x} \geq 102.33$ when $\mu = 99$)

$= P\left(z \geq \dfrac{102 - 99}{1}\right) = P(z \geq 3.33) = .0004$. Similarly,

$\alpha(98) = P(\bar{x} \geq 102.33$ when $\mu = 98) = P(z \geq 4.33) = 0$. In general, we have P(type I error) $< .01$ when this probability is calculated for a value of μ less than 100. The boundary value $\mu = 100$ yields the largest α.

Section 8.2

15.

a. $\alpha = P(z \geq 1.88$ when z has a standard normal distribution)
$= 1 - \Phi(1.88) = .0301$

b. $\alpha = P(z \leq -2.75$ when $z \sim N(0, 1) = \Phi(-2.75) = .003$

c. $\alpha = \Phi(-2.88) + (1 - \Phi(2.88)) = .004$

17.

a. $z = \dfrac{20{,}960 - 20{,}000}{1500 \Big/ \sqrt{16}} = 2.56 > 2.33$ so reject H_o.

b. $\beta(20{,}500): \Phi\left(2.33 + \dfrac{20{,}000 - 20{,}500}{1500/\sqrt{16}}\right) = \Phi(1.00) = .8413$

c. $\beta(20{,}500) = .05 : n = \left[\dfrac{1500(2.33 + 1.645)}{20{,}000 - 20{,}500}\right]^2 = 142.2$, so use n = 143

d. $\alpha = 1 - \Phi(2.56) = .0052$

19.

a. Reject H_o if either $z \ge 2.58$ or $z \le -2.58$; $\dfrac{\sigma}{\sqrt{n}} = 0.3$, so

$z = \dfrac{94.32 - 95}{0.3} = -2.27$. Since -2.27 is not < -2.58, don't reject H_o.

b. $\beta(94) = \Phi\left(2.58 - \dfrac{1}{0.3}\right) - \Phi\left(-2.58 - \dfrac{1}{0.3}\right) = \Phi(-.75) - \Phi(-5.91) = .2266$

c. $n = \left[\dfrac{1.20(2.58 + 1.28)}{95 - 94}\right]^2 = 21.46$, so use n = 22.

21. With H_o: $\mu = .5$, and H_a: $\mu \ne .5$ we reject H_o if $t > t_{\alpha/2, n-1}$ or $t < -t_{\alpha/2, n-1}$

a. $1.6 < t_{.025, 12} = 2.179$, so don't reject H_o

b. $-1.6 > -t_{.025, 12} = -2.179$, so don't reject H_o

c. $-2.6 > -t_{.005, 24} = -2.797$, so don't reject H_o

d. $-3.9 <$ the negative of all t values in the df = 24 row, so we reject H_o in favor of H_a.

23. H_o: $\mu = 360$ vs. H_a: $\mu > 360$; $t = \dfrac{\bar{x} - 360}{s / \sqrt{n}}$; reject H_o if $t > t_{.05,25} = 1.708$;

$t = \dfrac{370.69 - 360}{24.36 / \sqrt{26}} = 2.24 > 1.708$. Thus H_o should be rejected. There appears

to be a contradiction of the prior belief.

25.

a. H_o: $\mu = 5.5$ vs. H_a: $\mu \neq 5.5$; for a level .01 test, (not specified in the problem description), reject H_o if either $z \geq 2.58$ or $z \leq -2.58$. Since

$z = \dfrac{5.25 - 5.5}{.075} = -3.33 \leq -2.58$, reject H_o.

b. $1 - \beta(5.6) = 1 - \Phi\left(2.58 + \dfrac{(-.1)}{.075}\right) + \Phi\left(-2.58 - \dfrac{(-.1)}{.075}\right)$

$= 1 - \Phi(1.25) + \Phi(-3.91) = .105$

c. $n = \left[\dfrac{.3(2.58 + 2.33)}{-.1}\right]^2 = 216.97$, so use n = 217.

27. We wish to test H_o: $\mu = 75$ vs. H_a: $\mu < 75$; Using $\alpha = .01$, H_o is rejected if $t \leq -t_{.01,41} \approx -2.423$ (from the df 40 row of the t-table). Since

$t = \dfrac{73.1 - 75}{5.9 / \sqrt{42}} = -2.09$, which is not ≤ -2.423, H_o is not rejected. The alloy is not suitable.

29.

a. For n = 8, n − 1 = 7, and $t_{.05,7} = 1.895$, so H_o is rejected at level .05 if

$t \geq 1.895$. Since $\dfrac{s}{\sqrt{n}} = \dfrac{1.25}{\sqrt{8}} = .442$, $t = \dfrac{3.72 - 3.50}{.442} = .498$; this does

not exceed 1.895, so H_o is not rejected.

b. $d = \dfrac{|\mu_o - \mu|}{\sigma} = \dfrac{|3.50 - 4.00|}{1.25} = .40$, and n = 8, so from table A.17,

$\beta(4.0) \approx .72$

31. The hypotheses of interest are H_o: $\mu = 7$ vs H_a: $\mu < 7$, so a lower-tailed test is appropriate; H_o should be rejected if $t \leq -t_{.1,8} = -1.397$.

$t = \dfrac{6.32 - 7}{1.65/\sqrt{9}} = -1.24$. Because -1.24 is not ≤ -1.397, H_o (prior belief) is not rejected (contradicted) at level .01.

33. $\beta(\mu_o - \Delta) = \Phi\left(z_{\alpha/2} + \Delta\sqrt{n}/\sigma\right) - \Phi\left(-z_{\alpha/2} - \Delta\sqrt{n}/\sigma\right)$

$= 1 - \left[\Phi\left(-z_{\alpha/2} - \Delta\sqrt{n}/\sigma\right) + \Phi\left(z_{\alpha/2} - \Delta\sqrt{n}/\sigma\right)\right] = \beta(\mu_o + \Delta)$

(since 1 - Φ(c) = Φ(-c)).

Section 8.3

35.

1   Parameter of interest: p = true proportion of cars in this particular county passing emissions testing on the first try.

2   H_o: $p = .70$

3   H_a: $p \neq .70$

4    $z = \dfrac{\hat{p} - p_o}{\sqrt{p_o(1 - p_o)/n}} = \dfrac{\hat{p} - .70}{\sqrt{.70(.30)/n}}$

5   either $z \geq 1.96$ or $z \leq -1.96$

6    $z = \dfrac{124/200 - .70}{\sqrt{.70(.30)/200}} = -2.469$

7   Reject H_o. The data indicates that the proportion of cars passing the first time on emission testing or this county differs from the proportion of cars passing statewide.

37.

1   p = true proportion of the population with type A blood

2   H_o: $p = .40$

3   H_a: $p \neq .40$

4    $z = \dfrac{\hat{p} - p_o}{\sqrt{p_o(1 - p_o)/n}} = \dfrac{\hat{p} - .40}{\sqrt{.40(.60)/n}}$

5   Reject H_o if $z \geq 2.58$ or $z \leq -2.58$

6    $z = \dfrac{82/150 - .40}{\sqrt{.40(.60)/150}} = \dfrac{.147}{.04} = 3.667$

7   Reject H_o. The data does suggest that the percentage of the population with type A blood differs from 40%. (at the .01 significance level). Since the z critical value for a significance level of .05 is less than that of .01, the conclusion would not change.

39. Let p denote the true proportion of those called to appear for service who are black. We wish to test H_o: p = .25 vs H_a: p < .25. We use $z = \dfrac{\hat{p} - .25}{\sqrt{.25(.75)/n}}$, with the

rejection region $z \le -z_{.01} = -2.33$. We calculate $\hat{p} = \dfrac{177}{1050} = .1686$, and

$z = \dfrac{.1686 - .25}{.0134} = -6.1$. Because $-6.1 < -2.33$, H_o is rejected. A conclusion that

discrimination exists is very compelling.

41.

a. The alternative of interest here is H_a: p > .50 (which states that more than 50% of all enthusiasts prefer gut), so the rejection region should consist of large values of X (an upper-tailed test). Thus (15, 16, 17, 18, 19, 20) is the appropriate region.

b. $\alpha = P(15 \le X$ when $p = .5) = 1 - B(14; 20, .05) = .021$, so this is a level .05 test. For R = {14, 15, ..., 20}, $\alpha = .058$, so this R does not specify a level .05 test and the region of **a** is the best level .05 test. ($\alpha \le .05$ along with smallest possible β).

c. $\beta(.6) = B(14; 20, .6) = .874$, and $\beta(.8) = B(14; 20, .8) = .196$.

d. The best level .10 test is specified by R = (14, ..., 20} (with $\alpha = .052$) Since 13 is not in R, H_o is not rejected at this level.

43. H_o: p = .035 vs H_a: p < .035. We use $z = \dfrac{\hat{p} - .035}{\sqrt{.035(.965)/n}}$, with the rejection

region $z \le -z_{.01} = -2.33$. With $\hat{p} = \dfrac{15}{500} = .03$, $z = \dfrac{-.005}{\sqrt{.0082}} = -.61$. Because -

.61 isn't ≤ -2.33, H_o is not rejected. Robots have not demonstrated their superiority.

Section 8.4

45.

    a. p-value = .084 > .05 = α, so don't reject H_o.

    b. p-value = .003 < .001 = α, so reject H_o.

    c. .498 >> .05, so H_o can't be rejected at level .05

    d. 084 < .10, so reject H_o at level .10

    e. .039 is not < .01, so don't reject H_o.

    f. p-value = .218 > .10, so H_o cannot be rejected.

47.

    a. .0358

    b. .0802

    c. .5824

    d. .1586

    e. 0

49. An upper-tailed test

    a. Df = 14, α=.05; $t_{.05,14} = 1.761 : 3.2 > 1.761$, so reject H_o.

    b. $t_{.01,18} = 2.896$; 1.8 is not > 2.896, so don't reject H_o.

    c. Df = 23, p-value > .50, so fail to reject H_o at any significance level.

51. Here we might be concerned with departures above as well as below the specified weight of 5.0, so the relevant hypotheses are H_o: $\mu = 5.0$ vs H_a: $\mu \neq 5.0$. At level .01, reject H_o if either $z \geq 2.58$ or $z \leq -2.58$. Since $\dfrac{s}{\sqrt{n}} = .035$,

$$z = \frac{-.13}{.035} = -3.71,$$ which is ≤ -2.58, so H_o should be rejected. Because 3.71 is "off" the z-table, p-value < 2(.0002) = .0004 (.0002 corresponds to z = -3.49).

53. p = proportion of all physicians that know the generic name for methadone.

H_0: $p = .50$ vs H_a: $p < .50$; We can use a large sample test if both $np_0 \geq 10$ and

$n(1 - p_0) \geq 10$; $102(.50) = .51$, so we can proceed. $\hat{p} = \frac{47}{102}$, so

$$z = \frac{\frac{47}{102} - .50}{\sqrt{\frac{(.50)(.50)}{102}}} = \frac{-.039}{.050} = -.79 .$$ We will reject H_0 if the p-value < .01. For this

lower tailed test, the p-value = $\Phi(z) = \Phi(-.79) = .2148$, which is not < .01, so we do not reject H_0 at significance level .01.

55.   The hypotheses to be tested are H_0: $\mu = 25$ vs H_a: $\mu > 25$, and H_0 should be

rejected if $t \geq t_{.05,12} = 1.782$. The computed summary statistics are $\bar{x} = 27.923$,

$s = 5.619$, so $\dfrac{s}{\sqrt{n}} = 1.559$ and $t = \dfrac{2.923}{1.559} = 1.88$. From table A.8, P(t > 1.88)

$\approx .041$, which is less than .05, so H_0 is rejected at level .05.

57. μ = true average reading, H_0: $\mu = 70$ vs H_a: $\mu \neq 70$, and

$$t = \frac{\bar{x} - 70}{s/\sqrt{n}} = \frac{75.5 - 70}{7/\sqrt{6}} = \frac{5.5}{2.86} = 1.92 .$$ From table A.8, df = 5, p-value =

$2[P(t > 1.92)] \approx 2(.058) = .116$. At significance level .05, there is not enough evidence to conclude that the spectrophotometer needs recalibrating.

Section 8.5

59.

   a. The formula for β is $1 - \Phi\left(-2.33 + \dfrac{\sqrt{n}}{9.4} \right)$, which gives .8980 for n = 100,

     .1049 for n = 900, and .0014 for n = 2500.

   b. Z = -5.3, which is "off the z table," so p-value < .0002; this value of z is quite statistically significant.

   c. No. Eve when the departure from H_0 is insignificant from a practical point of view, a statistically significant result is highly likely to appear; the test is too likely to detect small departures from H_0.

Supplementary Exercises

61. Because n = 50 is large, we use a z test here, rejecting H_o: $\mu = 3.2$ in favor of H_a: $\mu \neq 3.2$ if either $z \geq z_{.025} = 1.96$ or $z \leq -1.96$. The computed z value is $z = \dfrac{3.05 - 3.20}{.34 / \sqrt{50}} = -3.12$. Since -3.12 is ≤ -1.96, H_o should be rejected in favor of H_a.

63.

    a. H_o: $\mu = 3.2$ vs H_a: $\mu \neq 3.2$ (Because H_a: $\mu > 3.2$ gives a p-value of roughly .15)

    b. With a p-value of .30, we would reject the null hypothesis at any reasonable significance level, which includes both .05 and .10.

65.

    a. The relevant hypotheses are H_o: $\mu = 548$ vs H_a: $\mu \neq 548$. At level .05, H_o will be rejected if either $t \geq t_{.025,10} = 2.228$ or $t \leq -t_{.025,10} = -2.228$. The test statistic value is $t = \dfrac{587 - 548}{10 / \sqrt{11}} = \dfrac{39}{3.02} = 12.9$. This clearly falls into the upper tail of the two-tailed rejection region, so H_o should be rejected at level .05, or any other reasonable level).

    b. The population sampled was normal or approximately normal.

Chapter 8: Tests of Hypotheses Based on a Single Sample

67. $N = 47$, $\bar{x} = 215$ mg, $s = 235$ mg. Range 5 mg to 1,176 mg.

   a. No, the distribution does not appear to be normal, it appears to be skewed to the right. It is not necessary to assume normality if the sample size is large enough due to the central limit theorem. This sample size is large enough so we can conduct a hypothesis test about the mean.

   b.

      1   Parameter of interest: μ = true daily caffeine consumption of adult women.

      2   H_o: $\mu = 200$

      3   H_a: $\mu > 200$

      4      $z = \dfrac{\bar{x} - 200}{s/\sqrt{n}}$

      5   RR: $z \geq 1.282$ or if p-value $\leq .10$

      6      $z = \dfrac{215 - 200}{235/\sqrt{47}} = .44$; p-value $= 1 - \Phi(.44) = .33$

      7   Fail to reject H_o. because $.33 > .10$. The data does not indicate that daily consumption of all adult women exceeds 200 mg.

69.

   a. From table A.17, when $\mu = 9.5$, $d = .625$, $df = 9$, and $\beta \approx .60$, when $\mu = 9.0$, $d = 1.25$, $df = 9$, and $\beta \approx .20$.

   b. From Table A.17, $\beta = .25$, $d = .625$, $n \approx 28$

71.

   a. With H_o: $p = \frac{1}{75}$ vs H_o: $p \neq \frac{1}{75}$, we reject H_o if either $z \geq 1.96$ or $z \leq -1.96$. With $\hat{p} = \dfrac{16}{800} = .02$, $z\dfrac{.02 - .01333}{\sqrt{\dfrac{.01333(.98667)}{800}}} = 1.645$, which

is not in either rejection region. Thus, we fail to reject the null hypothesis. There is not evidence that the incidence rate among prisoners differs from that of the adult population. The possible error we could have made is a type II.

   b. P-value $= 2[1 - \Phi(1.645)] = 2[.05] = .10$. Yes, since $.10 < .20$, we could reject H_o.

Chapter 8: Tests of Hypotheses Based on a Single Sample

73. Even though the underlying distribution may not be normal, a z test can be used because n is large. H_o: $\mu = 3200$ should be rejected in favor of H_a: $\mu < 3200$ if $z \le -z_{.001} = -3.08$. The computed z is $z = \dfrac{3107 - 3200}{188/\sqrt{45}} = -3.32 \le -3.08$, so H_o should be rejected at level .001.

75. We wish to test H_o: $\lambda = 4$ vs H_a: $\lambda > 4$ using the test statistic $z = \dfrac{\bar{x} - 4}{\sqrt{4/n}}$. For the given sample, n = 36 and $\bar{x} = \dfrac{160}{36} = 4.444$, so $z = \dfrac{4.444 - 4}{\sqrt{4/36}} = 1.33$. At level .02, we reject H_o if $z \ge z_{.02} \doteq 2.05$ (since $1 - \Phi(2.05) = .0202$). Because 1.33 is not ≥ 2.05, H_o should not be rejected at this level.

77. H_o: $\sigma^2 = .25$ vs H_a: $\sigma^2 > .25$. The chi-squared critical value for 9 d.f. that captures upper-tail area .01 is 21.665. The test statistic value is $\dfrac{9(.58)^2}{.25} = 12.11$. Because 12.11 is not ≥ 21.665, H_o cannot be rejected. The uniformity specification is not contradicted.

79.

a. $E(\bar{X} + 2.33S) = E(\bar{X}) + 2.33E(S) = \mu + 2.33\sigma$, so $\hat{\theta} = \bar{X} + 2.33S$ is approximately unbiased.

b. $V(\bar{X} + 2.33S) = V(\bar{X}) + 2.33^2 V(S) = \dfrac{\sigma^2}{n} + 5.4289\dfrac{\sigma^2}{2n}$. The estimated standard error (standard deviation) is $1.927\dfrac{s}{\sqrt{n}}$.

c. More than 99% of all soil samples have pH less than 6.75 iff the 95th percentile is less than 6.75. Thus we wish to test H$_o$: $\mu + 2.33\sigma = 6.75$ vs H$_a$: $\mu + 2.33\sigma < 6.75$. H$_o$ will be rejected at level .01 if $z \leq 2.33$. Since

$$z = \frac{-.047}{.0385} < 0,$$ H$_o$ clearly cannot be rejected. The 95th percentile does not appear to exceed 6.75.

81.

a. P(type I error) = P(either $Z \geq z_\gamma$ or $Z \leq z_{\alpha-\gamma}$) (when Z is a standard normal r.v.) = $\Phi(-z_{\alpha-\gamma}) + 1 - \Phi(z_\gamma) = \alpha - \gamma + \gamma = \alpha$.

b. $\beta(\mu) = P(\overline{X} \geq \mu_o + \frac{\sigma z_\gamma}{\sqrt{n}} \ or \overline{X} \leq \mu_o - \frac{\sigma z_{\alpha-\gamma}}{\sqrt{n}}$ when the true value is μ) =

$$\Phi\left(z_\gamma + \frac{\mu_o - \mu}{\sigma/\sqrt{n}}\right) - \Phi\left(-z_{\alpha-\gamma} + \frac{\mu_o - \mu}{\sigma/\sqrt{n}}\right)$$

c. Let $\lambda = \sqrt{n}\frac{\Delta}{\sigma}$; then we wish to know when $\pi(\mu_o + \Delta) = 1 - \Phi(z_\gamma - \lambda)$
$+ \Phi(-z_{\alpha-\gamma} - \lambda) > 1 - \Phi(z_\gamma + \lambda) + \Phi(-z_{\alpha-\gamma} + \lambda) = \pi(\mu_o - \Delta)$. Using the fact that $\Phi(-c) = 1 - \Phi(c)$, this inequality becomes
$\Phi(z_\gamma + \lambda) - \Phi(z_\gamma - \lambda) > \Phi(z_{\alpha-\gamma} + \lambda) - \Phi(z_{\alpha-\gamma} - \lambda)$. The l.h.s. is the area under the Z curve above the interval $(z_\gamma + \lambda, z_\gamma - \lambda)$, while the r.h.s. is the area above $(z_{\alpha-\gamma} - \lambda, z_{\alpha-\gamma} + \lambda)$. Both intervals have width 2λ, but when $z_\gamma < z_{\alpha-\gamma}$, the first interval is closer to 0 (and thus corresponds to the large area) than is the second. This happens when $\gamma > \alpha - \gamma$, i.e., when $\gamma > \alpha/2$.

CHAPTER 9

Section 9.1

1.

    a. $E(\overline{X} - \overline{Y}) = E(\overline{X}) - E(\overline{Y}) = 4.1 - 4.5 = -.4$, irrespective of sample sizes.

    b. $V(\overline{X} - \overline{Y}) = V(\overline{X}) + V(\overline{Y}) = \dfrac{\sigma_1^2}{m} + \dfrac{\sigma_2^2}{n} = \dfrac{(1.8)^2}{100} + \dfrac{(2.0)^2}{100} = .0724$, and

    the s.d. of $\overline{X} - \overline{Y} = \sqrt{.0724} = .2691$.

    c.   A normal curve with mean and s.d. as given in **a** and **b** (because m = n = 100, the CLT implies that both $\overline{X}$ and $\overline{Y}$ have approximately normal distributions, so $\overline{X} - \overline{Y}$ does also). The shape is not necessarily that of a normal curve when m = n = 10, because the CLT cannot be invoked. So if the two lifetime population distributions are not normal, the distribution of $\overline{X} - \overline{Y}$ will typically be quite complicated.

3.   The test statistic value is $z = \dfrac{(\overline{x} - \overline{y}) - 5000}{\sqrt{\dfrac{s_1^2}{m} + \dfrac{s_2^2}{n}}}$, and H$_o$ will be rejected at level .01 if

$z \geq 2.33$. We compute $z = \dfrac{(43{,}500 - 36{,}800) - 5000}{\sqrt{\dfrac{2200^2}{45} + \dfrac{1500^2}{45}}} = \dfrac{700}{396.93} = 1.76$,

which is not > 2.33, so we don't reject H$_o$ and conclude that the true average life for radials does not exceed that for economy brand by more than 500.

5.

a. H_a says that the average calorie output for sufferers is more than 1 cal/cm^2/min below that for nonsufferers. $\sqrt{\dfrac{\sigma_1^2}{m}+\dfrac{\sigma_2^2}{n}}=\sqrt{\dfrac{(.04)^2}{10}+\dfrac{(.16)^2}{10}}=.1414$, so

$z=\dfrac{(.64-2.05)-(-1)}{.1414}=-2.90$. At level .01, H_o is rejected if $z\le-2.33$; since $-2.90 < -2.33$, reject H_o.

b. $P=\Phi(-2.90)=.0019$

c. $\beta=1-\Phi\left(-2.33-\dfrac{-1.2+1}{.1414}\right)=1-\Phi(-.92)=.8212$

d. $m=n=\dfrac{.2(2.33+1.28)^2}{(-.2)^2}=65.15$, so use 66.

7.

1   Parameter of interest: $\mu_1-\mu_2=$ the true difference of means for males and females on the Boredom Proneness Rating. Let $\mu_1=$ men's average and $\mu_2=$ women's average.

2     $H_o: \mu_1-\mu_2=0$

3     $H_a: \mu_1-\mu_2>0$

4     $z=\dfrac{(\bar{x}-\bar{y})-\Delta_o}{\sqrt{\dfrac{s_1^2}{m}+\dfrac{s_2^2}{n}}}=\dfrac{(\bar{x}-\bar{y})-0}{\sqrt{\dfrac{s_1^2}{m}+\dfrac{s_2^2}{n}}}$

5   RR: $z\ge1.645$

6     $z=\dfrac{(10.40-9.26)-\Delta_o}{\sqrt{\dfrac{4.83^2}{97}+\dfrac{4.68^2}{148}}}=1.83$

7   Reject H_o. The data indicates the Boredom Proneness Rating is higher for males than for females.

9.

a. point estimate $\bar{x} - \bar{y} = 19.9 - 13.7 = 6.2$. It appears that there could be a difference.

b.

H_o: $\mu_1 - \mu_2 = 0$, H_a: $\mu_1 - \mu_2 \neq 0$,

$$z = \frac{(19.9 - 13.7)}{\sqrt{\dfrac{39.1^2}{60} + \dfrac{15.8^2}{60}}} = \frac{6.2}{5.44} = 1.14 \text{, and the p-value} = 2[P(z > 1.14)] = 2($$

.1271) = .2542. The p value is larger than any reasonable α, so we do not reject H_0. There is no significant difference.

c. No. With a normal distribution, we would expect most of the data to be within 2 standard deviations of the mean, and the distribution should be symmetric. 2 sd's above the mean is 98.1, but the distribution stops at zero on the left. The distribution is positively skewed.

d. We will calculate a 95% confidence interval for μ, the true average length of stays for patients given the treatment.

$$19.9 \pm 1.96 \frac{39.1}{\sqrt{60}} = 19.9 \pm 9.9 = (10.0, 21.8)$$

11. $(\bar{X} - \bar{Y}) \pm z_{\alpha/2} \sqrt{\dfrac{s_1^2}{m} + \dfrac{s_2^2}{n}}$. Standard error $= \dfrac{s}{\sqrt{n}}$. Substitution yields

$(\bar{x} - \bar{y}) \pm z_{\alpha/2} \sqrt{(SE_1)^2 + (SE_2)^2}$. Using $\alpha = .05$, $z_{\alpha/2} = 1.96$, so

$(5.5 - 3.8) \pm 1.96 \sqrt{(0.3)^2 + (0.2)^2} = (0.99, 2.41)$. Because we selected

$\alpha = .05$, we can state that when using this method with repeated sampling, the interval calculated will bracket the true difference 95% of the time. The interval is fairly narrow, indicating precision of the estimate.

13. $\sigma_1 = \sigma_2 = .05$, d = .04, $\alpha = .01$, $\beta = .05$, and the test is one-tailed, so

$$n = \frac{(.0025 + .0025)(2.33 + 1.645)^2}{.0016} = 49.38 \text{, so use n} = 50.$$

15.

a. As either m or n increases, σ decreases, so $\dfrac{\mu_1 - \mu_2 - \Delta_o}{\sigma}$ increases (the numerator is positive), so $\left(z_\alpha - \dfrac{\mu_1 - \mu_2 - \Delta_o}{\sigma} \right)$ decreases, so

$$\beta = \Phi\left(z_\alpha - \frac{\mu_1 - \mu_2 - \Delta_o}{\sigma} \right) \text{ decreases.}$$

b. As β decreases, z_β increases, and since z_β is the numerator of n, n increases also.

Section 9.2

17.

a. $v = \dfrac{\left(\frac{5^2}{10} + \frac{6^2}{10} \right)^2}{\dfrac{\left(\frac{5^2}{10} \right)^2}{9} + \dfrac{\left(\frac{6^2}{10} \right)^2}{9}} = \dfrac{37.21}{.694 + 1.44} = 17.43 \approx 17$

b. $v = \dfrac{\left(\frac{5^2}{10} + \frac{6^2}{15} \right)^2}{\dfrac{\left(\frac{5^2}{10} \right)^2}{9} + \dfrac{\left(\frac{6^2}{15} \right)^2}{14}} = \dfrac{24.01}{.694 + .411} = 21.7 \approx 21$

c. $v = \dfrac{\left(\frac{2^2}{10} + \frac{6^2}{15} \right)^2}{\dfrac{\left(\frac{2^2}{10} \right)^2}{9} + \dfrac{\left(\frac{6^2}{15} \right)^2}{14}} = \dfrac{7.84}{.018 + .411} = 18.27 \approx 18$

d. $v = \dfrac{\left(\frac{5^2}{12} + \frac{6^2}{24} \right)^2}{\dfrac{\left(\frac{5^2}{12} \right)^2}{11} + \dfrac{\left(\frac{6^2}{24} \right)^2}{23}} = \dfrac{12.84}{.395 + .098} = 26.05 \approx 26$

Chapter 9: Inferences Based on Two Samples

19.   For the given hypotheses, the test statistic
$$t = \frac{115.7 - 129.3 + 10}{\sqrt{\frac{5.03^2}{6} + \frac{5.38^2}{6}}} = \frac{-3.6}{3.007} = -1.20 \text{, and the d.f. is}$$

$$v = \frac{(4.2168 + 4.8241)^2}{\frac{(4.2168)^2}{5} + \frac{(4.8241)^2}{5}} = 9.96 \text{, so use d.f.} = 9. \text{ We will reject } H_o \text{ if}$$

$t \le -t_{.01,9} = -2.764$; since $-1.20 > -2.764$, we don't reject H_o.

21.   Let μ_1 = the true average gap detection threshold for normal subjects, and μ_2 = the corresponding value for CTS subjects. The relevant hypotheses are H_o:
$\mu_1 - \mu_2 = 0$ vs. H_a: $\mu_1 - \mu_2 < 0$, and the test statistic

$$t = \frac{1.71 - 2.53}{\sqrt{.0351125 + .07569}} = \frac{-.82}{.3329} = -2.46. \text{ Using d.f.}$$

$$v = \frac{(.0351125 + .07569)^2}{\frac{(.0351125)^2}{7} + \frac{(.07569)^2}{9}} = 15.1 \text{, or 15, the rejection region is}$$

$t \le -t_{.01,15} = -2.602$. Since -2.46 is not ≤ -2.602, we fail to reject H_o. We have insufficient evidence to claim that the true average gap detection threshold for CTS subjects exceeds that for normal subjects.

23.

    a.   Normal plots

Normal Probability Plot for Poor Quality Fabric

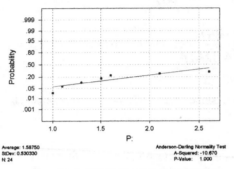

Average: 1.58750
StDev: 0.530330
N: 24

Anderson-Darling Normality Test
A-Squared: -10.670
P-Value: 1.000

Normal Probability Plot for High Quality Fabric

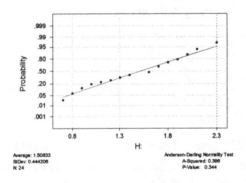

Average: 1.50833
StDev: 0.444206
N: 24

Anderson-Darling Normality Test
A-Squared: 0.396
P-Value: 0.344

We see that both plots illustrate sufficient linearity. Therefore, it is plausible that both samples have been selected from normal population distributions.

b.

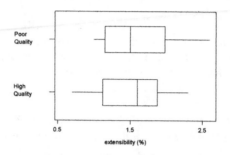

Comparative Box Plot for High Quality and Poor Quality Fabric

The comparative boxplot does not suggest a difference between average
extensibility for the two types of fabrics.

c.   We test $H_0 : \mu_1 - \mu_2 = 0$ vs. $H_a : \mu_1 - \mu_2 \neq 0$. With degrees of freedom

$$v = \frac{(.0433265)^2}{.00017906} = 10.5,$$ which we round down to 10, and using significance

level .05 (not specified in the problem), we reject H_o if $|t| \geq t_{.025,10} = 2.228$.

The test statistic is $t = \dfrac{-.08}{\sqrt{(.0433265)}} = -.38$, which is not ≥ 2.228 in

absolute value, so we cannot reject H_o. There is insufficient evidence to claim
that the true average extensibility differs for the two types of fabrics.

25.   We calculate the degrees of freedom $\nu = \dfrac{\left(\frac{5.5^2}{28} + \frac{7.8^2}{31}\right)^2}{\frac{\left(\frac{5.5^2}{28}\right)^2}{27} + \frac{\left(\frac{7.8^2}{31}\right)^2}{30}} = 53.95$, or about 54

(normally we would round down to 53, but this number is very close to 54 – of course for this large number of df, using either 53 or 54 won't make much difference in the critical t value) so the desired confidence interval is

$$(91.5 - 88.3) \pm 1.68\sqrt{\tfrac{5.5^2}{28} + \tfrac{7.8^2}{31}} = 3.2 \pm 2.931 = (.269, 6.131).$$ Because 0

does not lie inside this interval, we can be reasonably certain that the true difference $\mu_1 - \mu_2$ is not 0 and, therefore, that the two population means are not equal. For a 95% interval, the t value increases to about 2.01 or so, which results in the interval 3.2 ± 3.506. Since this interval does contain 0, we can no longer conclude that the means are different if we use a 95% confidence interval.

27.   The approximate degrees of freedom for this estimate are

$$\nu = \dfrac{\left(\frac{11.3^2}{6} + \frac{8.3^2}{8}\right)^2}{\frac{\left(\frac{11.3^2}{6}\right)^2}{5} + \frac{\left(\frac{8.3^2}{8}\right)^2}{7}} = \dfrac{893.59}{101.175} = 8.83$$, which we round down to 8, so

$t_{.025,8} = 2.306$ and the desired interval is

$$(40.3 - 21.4) \pm 2.306\sqrt{\tfrac{11.3^2}{6} + \tfrac{8.3^2}{8}} = 18.9 \pm 2.306(5.4674)$$

$= 18.9 \pm 12.607 = (6.3, 31.5).$ Because 0 is not contained in this interval, there is strong evidence that $\mu_1 - \mu_2$ is not 0; i.e., we can conclude that the population means are not equal. Calculating a confidence interval for $\mu_2 - \mu_1$ would change only the order of subtraction of the sample means, but the standard error calculation would give the same result as before. Therefore, the 95% interval estimate of $\mu_2 - \mu_1$ would be (-31.5, -6.3), just the negatives of the endpoints of the original interval. Since 0 is not in this interval, we reach exactly the same conclusion as before; the population means are not equal.

Chapter 9: Inferences Based on Two Samples

29. Let μ_1 = the true average compression strength for strawberry drink and let μ_2 = the true average compression strength for cola. A lower tailed test is appropriate. We test $H_0 : \mu_1 - \mu_2 = 0$ vs. $H_a : \mu_1 - \mu_2 < 0$. The test statistic is

$$t = \frac{-14}{\sqrt{29.4 + 15}} = -2.10 \, . \quad \nu = \frac{(44.4)^2}{\frac{(29.4)^2}{14} + \frac{(15)^2}{14}} = \frac{1971.36}{77.8114} = 25.3 \, , \text{ so use}$$

df=25. The p-value $\approx P(t < -2.10) = .023$. This p-value indicates strong support for the alternative hypothesis. The data does suggest that the extra carbonation of cola results in a higher average compression strength.

31.

a.

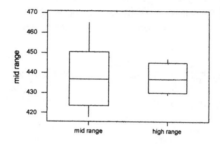

Comparative Box Plot for High Range and Mid Range

The most notable feature of these boxplots is the larger amount of variation present in the mid-range data compared to the high-range data. Otherwise, both look reasonably symmetric with no outliers present.

b. Using df = 23, a 95% confidence interval for $\mu_{mid-range} - \mu_{high-range}$ is

$$(438.3 - 437.45) \pm 2.069 \sqrt{\frac{15.1^2}{17} + \frac{6.83^2}{11}} = .85 \pm 8.69 = (-7.84, 9.54).$$

Since plausible values for $\mu_{mid-range} - \mu_{high-range}$ are both positive and negative (i.e., the interval spans zero) we would conclude that there is not sufficient evidence to suggest that the average value for mid-range and the average value for high-range differ.

33. Let μ_1 = the true average weight gain for steroid treatment and let μ_2 = the true average weight gain for the population not treated with steroids. The exercise asks if we can conclude that μ_2 exceeds μ_1 by more than 5 g., which we can restate in the equivalent form: $\mu_1 - \mu_2 < -5$. Therefore, we conduct a lower-tailed test of $H_0 : \mu_1 - \mu_2 = -5$ vs. $H_a : \mu_1 - \mu_2 < -5$. The test statistic is

$$t = \frac{(\bar{x} - \bar{y}) - (\Delta)}{\sqrt{\dfrac{s_1^2}{m} + \dfrac{s_2^2}{n}}} = \frac{32.8 - 40.5 - (-5)}{\sqrt{\dfrac{2.6^2}{8} + \dfrac{2.5^2}{10}}} = \frac{-2.7}{1.2124} = -2.23 \approx 2.2 \; . \text{ The approximate}$$

d.f. is $\nu = \dfrac{\left(\dfrac{2.6^2}{8} + \dfrac{2.5^2}{10}\right)^2}{\dfrac{\left(\dfrac{2.6^2}{8}\right)^2}{7} + \dfrac{\left(\dfrac{2.5^2}{10}\right)^2}{9}} = \dfrac{2.1609}{.1454} = 14.876$, which we round down to 14.

The p-value for a lower tailed test is $P(t < -2.2) = P(t > 2.2) = .022$. Since this p-value is larger than the specified significance level .01, we cannot reject H_0. Therefore, this data does not support the belief that average weight gain for the control group exceeds that of the steroid group by more than 5 g.

35. There are two changes that must be made to the procedure we currently use. First, the equation used to compute the value of the t test statistic is: $t = \dfrac{(\bar{x} - \bar{y}) - (\Delta)}{s_p \sqrt{\dfrac{1}{m} + \dfrac{1}{n}}}$ where

s_p is defined as in Exercise 34 above. Second, the degrees of freedom = $m + n - 2$. Assuming equal variances in the situation from Exercise 33, we calculate s_p as

follows: $s_p = \sqrt{\left(\dfrac{7}{16}\right)(2.6)^2 + \left(\dfrac{9}{16}\right)(2.5)^2} = 2.544$. The value of the test

statistic is, then, $t = \dfrac{(32.8 - 40.5) - (-5)}{2.544\sqrt{\dfrac{1}{8} + \dfrac{1}{10}}} = -2.24 \approx -2.2$. The degrees of

freedom = 16, and the p-value is $P(t < -2.2) = .021$. Since $.021 > .01$, we fail to reject H_0. This is the same conclusion reached in Exercise 33.

Section 9.3

37.

    a.   This exercise calls for paired analysis. First, compute the difference between indoor and outdoor concentrations of hexavalent chromium for each of the 33 houses. These 33 differences are summarized as follows: $n = 33$, $\bar{d} = -.4239$, $s_d = .3868$, where d = (indoor value – outdoor value). Then $t_{.025,32} = 2.037$, and a 95% confidence interval for the population mean difference between indoor and outdoor concentration is

$$-.4239 \pm (2.037)\left(\frac{.3868}{\sqrt{33}}\right) = -.4239 \pm .13715 = (-.5611, -.2868).$$ We

can be highly confident, at the 95% confidence level, that the true average concentration of hexavalent chromium outdoors exceeds the true average concentration indoors by between .2868 and .5611 nanograms/m^3.

    b.   A 95% prediction interval for the difference in concentration for the 34th house is

$$\bar{d} \pm t_{.025,32}\left(s_d\sqrt{1+\tfrac{1}{n}}\right) = -.4239 \pm (2.037)(.3868\sqrt{1+\tfrac{1}{33}}) = (-1.224, .3758)$$

. This prediction interval means that the indoor concentration may exceed the outdoor concentration by as much as .3758 nanograms/m^3 and that the outdoor concentration may exceed the indoor concentration by a much as 1.224 nanograms/m^3, for the 34th house. Clearly, this is a wide prediction interval, largely because of the amount of variation in the differences.

39.

    a.   A normal probability plot shows that the data could easily follow a normal distribution.

    b.   We test $H_0 : \mu_d = 0$ vs. $H_a : \mu_d \neq 0$, with test statistic

$$t = \frac{\bar{d} - 0}{s_D/\sqrt{n}} = \frac{167.2 - 0}{228/\sqrt{14}} = 2.74 \approx 2.7.$$ The two-tailed p-value is 2[P(t >

2.7)] = 2[.009] = .018. Since .018 < .05, we can reject H$_o$. There is strong evidence to support the claim that the true average difference between intake values measured by the two methods is not 0. There is a difference between them.

41. We test $H_0 : \mu_d = 0$ vs. $H_a : \mu_d > 0$. With $\overline{d} = 7.600$, and $s_d = 4.178$,

$$t = \frac{7.600 - 5}{4.178 / \sqrt{9}} = \frac{2.6}{1.39} = 1.87 \approx 1.9.$$ With degrees of freedom $n - 1 = 8$, the

corresponding p-value is $P(t > 1.9) = .047$. We would reject H_0 at any alpha level greater than .047. So, at the typical significance level of .05, we would (barely) reject H_0, and conclude that the data indicates that the higher level of illumination yields a decrease of more than 5 seconds in true average task completion time.

43.

   a. Although there is a "jump" in the middle of the Normal Probability plot, the data follow a reasonably straight path, so there is no strong reason for doubting the normality of the population of differences.

   b. A 95% lower confidence bound for the population mean difference is:

$$\overline{d} - t_{.05,14} \left(\frac{s_d}{\sqrt{n}} \right) = -38.60 - (1.761) \left(\frac{23.18}{\sqrt{15}} \right) = -38.60 - 10.54 = -49.14$$

. Therefore, with a confidence level of 95%, the population mean difference is above (–49.14).

   c. A 95% upper confidence bound for the corresponding population mean difference is $38.60 + 10.54 = 49.14$

45. The differences (white – black) are –7.62, -8.00, -9.09, -6.06, -1.39, -16.07, -8.40, -8.89, and –2.88, from which $\overline{d} = -7.600$, and $s_d = 4.178$. The confidence level is not specified in the problem description; for 95% confidence, $t_{.025,8} = 2.306$, and the C.I. is

$$-7.600 \pm (2.306) \left(\frac{4.178}{\sqrt{9}} \right) = -7.600 \pm 3.211 = (-10.811, -4.389).$$

Section 9.4

47. H_o will be rejected if $z \le -z_{.01} = -2.33$. With $\hat{p}_1 = .150$, and $\hat{p}_2 = .300$,

$$\hat{p} = \frac{30+80}{200+600} = \frac{210}{800} = .263, \text{ and } \hat{q} = .737. \text{ The calculated test statistic is}$$

$$z = \frac{.150-.300}{\sqrt{(.263)(.737)\left(\frac{1}{200}+\frac{1}{600}\right)}} = \frac{-.150}{.0359} = -4.18. \text{ Because } -4.18 \le -2.33,$$

H_o is rejected; the proportion of those who repeat after inducement appears lower than those who repeat after no inducement.

49.

1   Parameter of interest: $p_1 - p_2$ = true difference in proportions of those responding to two different survey covers. Let p_1 = Plain, p_2 = Picture.

2      $H_0 : p_1 - p_2 = 0$

3      $H_a : p_1 - p_2 < 0$

4      $z = \dfrac{\hat{p}_1 - \hat{p}_2}{\sqrt{\hat{p}\hat{q}\left(\frac{1}{m}+\frac{1}{n}\right)}}$

5   Reject H_o if p-value $< .10$

6      $z = \dfrac{\frac{104}{207} - \frac{109}{213}}{\sqrt{\left(\frac{213}{420}\right)\left(\frac{207}{420}\right)\left(\frac{1}{207}+\frac{1}{213}\right)}} = -.1910$; p-value = .4247

7   Fail to Reject H_o. The data does not indicate that plain cover surveys have a lower response rate.

51.

a. $H_0 : p_1 = p_2$ will be rejected in favor of $H_a : p_1 \neq p_2$ if either $z \geq 1.645$ or $z \leq -1.645$. With $\hat{p}_1 = .193$, and $\hat{p}_2 = .182$, $\hat{p} = .188$,

$$z = \frac{.011}{.00742} = 1.48.$$ Since 1.48 is not ≥ 1.645, H_0 is not rejected and we conclude that no difference exists.

b. Using formula (9.7) with $p_1 = .2$, $p_2 = .18$, $\alpha = .1$, $\beta = .1$, and $z_{\alpha/2} = 1.645$,

$$n = \frac{\left(1.645\sqrt{.5(.38)(1.62)} + 1.28\sqrt{.16 + .1476}\right)^2}{.0004} = 6582$$

53.

a. A 95% large sample confidence interval formula for $\ln(\theta)$ is

$$\ln(\hat{\theta}) \pm z_{\alpha/2}\sqrt{\frac{m-x}{mx} + \frac{n-y}{ny}}.$$ Taking the antilogs of the upper and lower bounds gives the confidence interval for θ itself.

b. $\hat{\theta} = \frac{\frac{189}{11,034}}{\frac{104}{11,037}} = 1.818$, $\ln(\hat{\theta}) = .598$, and the standard deviation is

$$\sqrt{\frac{10,845}{(11,034)(189)} + \frac{10,933}{(11,037)(104)}} = .1213,$$ so the CI for $\ln(\theta)$ is $.598 \pm 1.96(.1213) = (.360, .836)$. Then taking the antilogs of the two bounds gives the CI for θ to be $(1.43, 2.31)$.

55. $\hat{p}_1 = \frac{15 + 7}{40} = .550$, $\hat{p}_2 = \frac{29}{42} = .690$, and the 95% C.I. is
$(.550 - .690) \pm 1.96(.106) = -.14 \pm .21 = (-.35, .07)$.

Section 9.5

57.

    a. From Table A.9, column 5, row 8, $F_{.01,5,8} = 3.69$.

    b. From column 8, row 5, $F_{.01,8,5} = 4.82$.

    c. $F_{.95,5,8} = \dfrac{1}{F_{.05,8,5}} = .207$.

    d. $F_{.95,8,5} = \dfrac{1}{F_{.05,5,8}} = .271$

    e. $F_{.01,10,12} = 4.30$

    f. $F_{.99,10,12} = \dfrac{1}{F_{.01,12,10}} = \dfrac{1}{4.71} = .212$.

    g. $F_{.05,6,4} = 6.16$, so $P(F \leq 6.16) = .95$.

    h. Since $F_{.99,10,5} = \dfrac{1}{5.64} = .177$,

        $P(.177 \leq F \leq 4.74) = P(F \leq 4.74) - P(F \leq .177) = .95 - .01 = .94$.

59.   We test $H_0 : \sigma_1^2 = \sigma_2^2$ vs. $H_a : \sigma_1^2 \neq \sigma_2^2$. The calculated test statistic is

$f = \dfrac{(2.75)^2}{(4.44)^2} = .384$. With numerator d.f. = m $-$ 1 = 10 $-$ 1 = 9, and denominator

d.f. = n $-$ 1 = 5 $-$ 1 = 4, we reject H$_0$ if $f \geq F_{.05,9,4} = 6.00$ or

$f \leq F_{.95,9,4} = \dfrac{1}{F_{.05,4,9}} = \dfrac{1}{3.63} = .275$. Since .384 is in neither rejection

region, we do not reject H$_0$ and conclude that there is no significant difference between the two standard deviations.

61. Let σ_1^2 = variance in weight gain for low-dose treatment, and σ_2^2 = variance in weight gain for control condition. We wish to test $H_0 : \sigma_1^2 = \sigma_2^2$ vs.

$H_a : \sigma_1^2 > \sigma_2^2$. The test statistic is $f = \dfrac{s_1^2}{s_2^2}$, and we reject H_o at level .05 if

$f > F_{.05,19,22} \approx 2.08$. $f = \dfrac{(54)^2}{(32)^2} = 2.85 \geq 20.8$, so reject H_o at level .05. The

data does suggest that there is more variability in the low-dose weight gains.

63. $P\left(F_{1-\alpha/2,m-1,n-1} \leq \dfrac{S_1^2/\sigma_1^2}{S_2^2/\sigma_2^2} \leq F_{\alpha/2,m-1,n-1} \right) = 1-\alpha$. The set of inequalities

inside the parentheses is clearly equivalent to

$\dfrac{S_2^2 F_{1-\alpha/2,m-1,n-1}}{S_1^2} \leq \dfrac{\sigma_2^2}{\sigma_1^2} \leq \dfrac{S_2^2 F_{\alpha/2,m-1,n-1}}{S_1^2}$. Substituting the sample values s_1^2 and

s_2^2 yields the confidence interval for $\dfrac{\sigma_2^2}{\sigma_1^2}$, and taking the square root of each

endpoint yields the confidence interval for $\dfrac{\sigma_2}{\sigma_1}$. m = n = 4, so we need

$F_{.05,3,3} = 9.28$ and $F_{.95,3,3} = \dfrac{1}{9.28} = .108$. Then with $s_1 = .160$ and $s_2 = .074$, the

C. I. for $\dfrac{\sigma_2^2}{\sigma_1^2}$ is (.023, 1.99), and for $\dfrac{\sigma_2}{\sigma_1}$ is (.15, 1.41).

Supplementary Exercises

65.   We test $H_0 : \mu_1 - \mu_2 = 0$ vs. $H_a : \mu_1 - \mu_2 \neq 0$. The test statistic is

$$t = \frac{(\bar{x} - \bar{y}) - (\Delta)}{\sqrt{\dfrac{s_1^2}{m} + \dfrac{s_2^2}{n}}} = \frac{807 - 757}{\sqrt{\dfrac{27^2}{10} + \dfrac{41^2}{10}}} = \frac{50}{\sqrt{241}} = \frac{50}{15.524} = 3.22. \text{ The}$$

approximate d.f. is $V = \dfrac{(241)^2}{\dfrac{(72.9)^2}{9} + \dfrac{(168.1)^2}{9}} = 15.6$, which we round down to 15.

The p-value for a two-tailed test is approximately 2P(t > 3.22) = 2(.003) = .006.
This small of a p-value gives strong support for the alternative hypothesis. The data
indicates a significant difference.

67.   Let p_1 = true proportion of returned questionnaires that included no incentive; p_2 =
true proportion of returned questionnaires that included an incentive. The hypotheses
are $H_0 : p_1 - p_2 = 0$ vs. $H_0 : p_1 - p_2 < 0$. The test statistic is

$$z = \frac{\hat{p}_1 - \hat{p}_2}{\sqrt{\hat{p}\hat{q}\left(\frac{1}{m} + \frac{1}{n}\right)}}. \quad \hat{p}_1 = \frac{75}{110} = .682, \text{ and } \hat{p}_2 = \frac{66}{98} = .673. \text{ At this point we}$$

notice that since $\hat{p}_1 > \hat{p}_2$, the numerator of the z statistic will be > 0, and since we
have a lower tailed test, the p-value will be > .5. We fail to reject H_o. This data does
not suggest that including an incentive increases the likelihood of a response.

69. The center of any confidence interval for $\mu_1 - \mu_2$ is always $\bar{x}_1 - \bar{x}_2$, so

$$\bar{x}_1 - \bar{x}_2 = \frac{-473.3 + 1691.9}{2} = 609.3.$$ Furthermore, half of the width of this

interval is $\dfrac{1691.9 - (-473.3)}{2} = 1082.6$. Equating this value to the expression

on the right of the 95% confidence interval formula, $1082.6 = (1.96)\sqrt{\dfrac{s_1^2}{n_1} + \dfrac{s_2^2}{n_2}}$,

we find $\sqrt{\dfrac{s_1^2}{n_1} + \dfrac{s_2^2}{n_2}} = \dfrac{1082.6}{1.96} = 552.35$. For a 90% interval, the associated z

value is 1.645, so the 90% confidence interval is then
$$609.3 \pm (1.645)(552.35) = 609.3 \pm 908.6 = (-299.3, 1517.9).$$

71. $m = n = 40$, $\bar{x} = 3975.0$, $s_1 = 245.1$, $\bar{y} = 2795.0$, $s_2 = 293.7$. The large sample

99% confidence interval for $\mu_1 - \mu_2$ is

$$(3975.0 - 2795.0) \pm 2.58\sqrt{\frac{245.1^2}{40} + \frac{293.7^2}{40}}$$

$(1180.0) \pm 1560.5 \approx (1024, 1336)$. The value 0 is not contained in this interval

so we can state that, with very high confidence, the value of $\mu_1 - \mu_2$ is not 0, which

is equivalent to concluding that the population means are not equal.

73.   Since we can assume that the distributions from which the samples were taken are normal, we use the two-sample t test. Let μ_1 denote the true mean headability rating for aluminum killed steel specimens and μ_2 denote the true mean headability rating for silicon killed steel. Then the hypotheses are $H_0 : \mu_1 - \mu_2 = 0$ vs. $H_a : \mu_1 - \mu_2 \neq 0$. The test statistic is

$$t = \frac{-.66}{\sqrt{.03888 + .047203}} = \frac{-.66}{\sqrt{.086083}} = -2.25.$$ The approximate degrees of

freedom $\nu = \dfrac{(.086083)^2}{\dfrac{(.03888)^2}{29} + \dfrac{(.047203)^2}{29}} = 57.5$, so we use 57. The two-tailed p-

value $\approx 2(.014) = .028$, which is less than the specified significance level, so we would reject H_0. The data supports the article's authors' claim.

75.

a.   The relevant hypotheses are $H_0 : \mu_1 - \mu_2 = 0$ vs. $H_a : \mu_1 - \mu_2 \neq 0$. Assuming both populations have normal distributions, the two-sample t test is appropriate. $m = 11$, $\bar{x} = 98.1$, $s_1 = 14.2$, $n = 15$, $\bar{y} = 129.2$, $s_2 = 39.1$. The

test statistic is $t = \dfrac{-31.1}{\sqrt{18.3309 + 101.9207}} = \dfrac{-31.1}{\sqrt{120.252}} = -2.84$. The

approximate degrees of freedom $\nu = \dfrac{(120.252)^2}{\dfrac{(18.3309)^2}{10} + \dfrac{(101.9207)^2}{14}} = 18.64$,

so we use 18. From Table A.7, the two-tailed p-value $\approx 2(.006) = .012$. No, obviously, the results are different.

b.   For the hypotheses $H_0 : \mu_1 - \mu_2 = -25$ vs. $H_a : \mu_1 - \mu_2 < -25$, the test

statistic changes to $t = \dfrac{-31.1 - (-25)}{\sqrt{120.252}} = -.556$. With degrees of freedom

18, the p-value $\approx P(t < -.6) = .278$. Since the p-value is greater than any sensible choice of α, we fail to reject H_0. There is insufficient evidence that the true average strength for males exceeds that for females by more than 25N.

77. This is paired data, so the paired t test is employed. The relevant hypotheses are $H_0 : \mu_d = 0$ vs. $H_a : \mu_d < 0$, where μ_d denotes the difference between the population average control strength minus the population average heated strength. The observed differences (control – heated) are: -.06, .01, -.02, 0, and -.05. The sample mean and standard deviation of the differences are $\overline{d} = -.024$ and $s_d = .0305$. The test statistic is $t = \dfrac{-.024}{.0305/\sqrt{5}} = -1.76 \approx -1.8$. From Table A.7, with d.f. = 5 – 1 = 4, the lower tailed p-value associated with t = -1.8 is P(t < -1.8) = P(t > 1.8) = .073. At significance level .05, H_0 should not be rejected. Therefore, this data does not show that the heated average strength exceeds the average strength for the control population.

79.

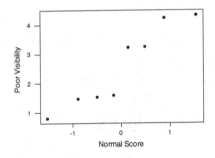

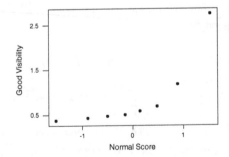

A normal probability plot indicates the data for good visibility does not follow a normal distribution, thus a t-test is not appropriate for this small a sample size.

81. We wish to test H_0: $\mu_1 = \mu_2$ versus H_a: $\mu_1 \neq \mu_2$

Unpooled:

With H_0: $\mu_1 - \mu_2 = 0$ vs. H_a: $\mu_1 - \mu_2 \neq 0$, we will reject H_o if

$$p-value < \alpha . \quad v = \frac{\left(\frac{.79^2}{14} + \frac{1.52^2}{12}\right)^2}{\frac{\left(\frac{.79^2}{14}\right)^2}{13} + \frac{\left(\frac{1.52^2}{12}\right)^2}{11}} = 15.95 \approx 16 \text{, and the test statistic}$$

$$t = \frac{8.48 - 9.36}{\sqrt{\frac{.79^2}{14} + \frac{1.52^2}{12}}} = \frac{-.96}{.4869} = -1.97 \text{ leads to a p-value of } 2[\, P(t > 1.97)]$$

$$\approx 2(.031) \approx .062$$

Pooled:

The degrees of freedom $v = m = n - 2 = 14 + 12 - 2 = 24$ and the pooled

variance is $\left(\frac{13}{24}\right)(.79)^2 + \left(\frac{11}{24}\right)(1.52)^2 = 1.3970$, so $s_p = 1.181$. The test

statistic is $t = \dfrac{-.96}{1.181\sqrt{\frac{1}{14} + \frac{1}{12}}} = \dfrac{-.96}{.465} \approx -2.1$. The p-value $= 2[\, P(t_{24} > 2.1\,)] =$

$2(.023) = .046$.

With the pooled method, there are more degrees of freedom, and the p-value is smaller than with the unpooled method.

83.

a. With n denoting the second sample size, the first is m = 3n. We then wish

$$20 = 2(2.58)\sqrt{\frac{900}{3n} + \frac{400}{n}} \text{, which yields n = 47, m = 141.}$$

b. We wish to find the n which minimizes $2(z_{\alpha/2})\sqrt{\dfrac{900}{400 - n} + \dfrac{400}{n}}$, or

equivalently, the n which minimizes $\dfrac{900}{400 - n} + \dfrac{400}{n}$. Taking the derivative

with respect to n and equating to 0 yields $900(400 - n)^{-2} - 400n^{-2} = 0$,

whence $9n^2 = 4(400 - n)^2$, or $5n^2 + 3200n - 640{,}000 = 0$. This yields n
= 160, m = 400 − n = 240.

85. Let p_1 = true survival rate at $11°C$; p_2 = true survival rate at $30°C$; The hypotheses are $H_0 : p_1 - p_2 = 0$ vs. $H_a : p_1 - p_2 \neq 0$. The test statistic is

$$z = \frac{\hat{p}_1 - \hat{p}_2}{\sqrt{\hat{p}\hat{q}(\frac{1}{m} + \frac{1}{n})}}. \text{ With } \hat{p}_1 = \frac{73}{91} = .802 \text{ , and } \hat{p}_2 = \frac{102}{110} = .927 \text{ ,}$$

$$\hat{p} = \frac{175}{201} = .871, \hat{q} = .129 \text{ . } z = \frac{.802 - .927}{\sqrt{(.871)(.129)(\frac{1}{91} + \frac{1}{110})}} = \frac{-.125}{.0320} = -3.91.$$

The p-value = $\Phi(-3.91) < \Phi(-3.49) = .0003$, so reject H_o at any reasonable level. The two survival rates appear to differ.

87. $\Delta_0 = 0$, $\sigma_1 = \sigma_2 = 10$, d = 1, $\sigma = \sqrt{\dfrac{200}{n}} = \dfrac{14.142}{\sqrt{n}}$, so

$$\beta = \Phi\left(1.645 - \frac{\sqrt{n}}{14.142}\right), \text{ giving } \beta = .9015, .8264, .0294, \text{ and } .0000 \text{ for n} = 25,$$

100, 2500, and 10,000 respectively. If the μ_i's referred to true average IQ's resulting from two different conditions, $\mu_1 - \mu_2 = 1$ would have little practical significance, yet very large sample sizes would yield statistical significance in this situation.

89. $H_0 : p_1 = p_2$ will be rejected at level α in favor of $H_a : p_1 > p_2$ if either $z \geq z_{.05} = 1.645$. With $\hat{p}_1 = \frac{250}{2500} = .10$, $\hat{p}_2 = \frac{167}{2500} = .0668$, and $\hat{p} = .0834$,

$$z = \frac{.0332}{.0079} = 4.2 \text{ , so } H_o \text{ is rejected . It appears that a response is more likely for a}$$

white name than for a black name.

91.

a. Let μ_1 and μ_2 denote the true average weights for operations 1 and 2, respectively. The relevant hypotheses are $H_0 : \mu_1 - \mu_2 = 0$ vs. $H_a : \mu_1 - \mu_2 \neq 0$. The value of the test statistic is

$$t = \frac{(1402.24 - 1419.63)}{\sqrt{\dfrac{(10.97)^2}{30} + \dfrac{(9.96)^2}{30}}} = \frac{-17.39}{\sqrt{4.011363 + 3.30672}} = \frac{-17.39}{\sqrt{7.318083}} = -6.43$$

. The d.f. $\nu = \dfrac{(7.318083)^2}{\dfrac{(4.011363)^2}{29} + \dfrac{(3.30672)^2}{29}} = 57.5$, so use df = 57.

$t_{.025,57} \approx 2.000$, so we can reject H_o at level .05. The data indicates that there is a significant difference between the true mean weights of the packages for the two operations.

b. $H_0 : \mu_1 = 1400$ will be tested against $H_a : \mu_1 > 1400$ using a one-sample t test with test statistic $t = \dfrac{\overline{x} - 1400}{s_1 / \sqrt{m}}$. With degrees of freedom = 29, we reject H_o if $t > t_{.05,29} = 1.699$. The test statistic value is

$$t = \frac{1402.24 - 1400}{10.97 / \sqrt{30}} = \frac{2.24}{2.00} = 1.1.$$ Because 1.1 < 1.699, H_o is not rejected.

True average weight does not appear to exceed 1400.

93. $\hat{\lambda}_1 = \overline{x} = 1.62$, $\hat{\lambda}_2 = \overline{y} = 2.56$, $\sqrt{\dfrac{\hat{\lambda}_1}{m} + \dfrac{\hat{\lambda}_2}{n}} = 1.77$, and the confidence interval is $-.94 \pm (1.96)(1.77) = -.94 \pm .35 = (-1.29, -.59)$

Chapter 9: Inferences Based on Two Samples

CHAPTER 10

Section 10.1

1.

    a. H_o will be rejected if $f \geq F_{.05,4,15} = 3.06$ (since $I - 1 = 4$, and $I(J-1) =$

    $(5)(3) = 15$). The computed value of F is $f = \dfrac{MSTr}{MSE} = \dfrac{2673.3}{1094.2} = 2.44$.

    Since 2.44 is not ≥ 3.06, H_o is not rejected. The data does not indicate a difference in the mean tensile strengths of the different types of copper wires.

    b. $F_{.05,4,15} = 3.06$ and $F_{.10,4,15} = 2.36$, and our computed value of 2.44 is between those values, it can be said that $.05 < $ p-value $ < .10$.

3.   With μ_i = true average lumen output for brand i bulbs, we wish to test

    $H_0 : \mu_1 = \mu_2 = \mu_3$ versus H_a : at least two $\mu_i 's$ are unequal.

    $MSTr = \hat{\sigma}_B^2 = \dfrac{591.2}{2} = 295.60$, $MSE = \hat{\sigma}_W^2 = \dfrac{4773.3}{21} = 227.30$, so

    $f = \dfrac{MSTr}{MSE} = \dfrac{295.60}{227.30} = 1.30$ For finding the p-value, we need degrees of

    freedom $I - 1 = 2$ and $I(J-1) = 21$. In the 2nd row and 21st column of Table A.9, we see that $1.30 < F_{.10,2,21} = 2.57$, so the p-value $> .10$. Since .10 is not $< .05$, we cannot reject H_o. There are no differences in the average lumen outputs among the three brands of bulbs.

5. μ_i = true mean modulus of elasticity for grade i (i = 1, 2, 3). We test

    $H_0 : \mu_1 = \mu_2 = \mu_3$ vs. H_a : at least two $\mu_i 's$ are unequal. Reject H_o if

    $f \geq F_{.01,2,27} = 5.49$. The grand mean = 1.5367,

    $MSTr = \dfrac{10}{2} \left[(1.63 - 1.5367)^2 + (1.56 - 1.5367)^2 + (1.42 - 1.5367)^2 \right] = .1143$

    $MSE = \dfrac{1}{3} \left[(.27)^2 + (.24)^2 + (.26)^2 \right] = .0660$, $f = \dfrac{MSTr}{MSE} = \dfrac{.1143}{.0660} = 1.73$.

    Fail to reject H_o. The three grades do not appear to differ.

7.

Source	Df	SS	MS	F
Treatments	3	75,081.72	25,027.24	1.70
Error	16	235,419.04	14,713.69	
Total	19	310,500.76		

The hypotheses are $H_0 : \mu_1 = \mu_2 = \mu_3 = \mu_4$ vs. H_a : at least two μ_i's are unequal. $1.70 < F_{.10,3,16} = 2.46$, so p-value $> .10$, and we fail to reject H_0.

9. The summary quantities are $x_{1\bullet} = 34.3$, $x_{2\bullet} = 39.6$, $x_{3\bullet} = 33.0$, $x_{4\bullet} = 41.9$,

$x_{\bullet\bullet} = 148.8$, $\Sigma\Sigma x_{ij}^2 = 946.68$, so $CF = \dfrac{(148.8)^2}{24} = 922.56$,

$SST = 946.68 - 922.56 = 24.12$,

$SSTr = \dfrac{(34.3)^2 + ... + (41.9)^2}{6} - 922.56 = 8.98$,

$SSE = 24.12 - 8.98 = 15.14$.

Source	Df	SS	MS	F
Treatments	3	8.98	2.99	3.95
Error	20	15.14	.757	
Total	23	24.12		

Since $3.10 = F_{.05,3,20} < 3.95 < 4.94 = F_{.01,3,20}$, $.01 < p - value < .05$ and H_0 is rejected at level .05.

Section 10.2

11. $Q_{.05,5,15} = 4.37$, $w = 4.37\sqrt{\dfrac{272.8}{4}} = 36.09$.

3	1	4	2	5
437.5	462.0	469.3	512.8	532.1

The brands seem to divide into two groups: 1, 3, and 4; and 2 and 5; with no significant differences within each group but all between group differences are significant.

13.

3	1	4	2	5
427.5	462.0	469.3	502.8	532.1

Brand 1 does not differ significantly from 3 or 4, 2 does not differ significantly from 4 or 5, 3 does not differ significantly from1, 4 does not differ significantly from 1 or 2, 5 does not differ significantly from 2, but all other differences (e.g., 1 with 2 and 5, 2 with 3, etc.) do appear to be significant.

15. $Q_{.01,4,36} = 4.75$, $w = 4.75\sqrt{\dfrac{15.64}{10}} = 5.94$.

2	1	3	4
24.69	26.08	29.95	33.84

Treatment 4 appears to differ significantly from both 1 and 2, but there are no other significant differences.

17. $\theta = \Sigma c_i \mu_i$ where $c_1 = c_2 = .5$ and $c_3 = -1$, so

$\hat{\theta} = .5\bar{x}_{1\bullet} + .5\bar{x}_{2\bullet} - \bar{x}_{3\bullet} = -.396$ and $\Sigma c_i^2 = 1.50$. With $t_{.025,6} = 2.447$ and MSE = .03106, the CI is (from 10.5 on page 418)

$$-.396 \pm (2.447)\sqrt{\frac{(.03106)(1.50)}{3}} = -.396 \pm .305 = (-.701, -.091).$$

19. MSTr = 140, error d.f. = 12, so $f = \dfrac{140}{SSE/12} = \dfrac{1680}{SSE}$ and $F_{.05,2,12} = 3.89$.

$w = Q_{.05,3,12}\sqrt{\dfrac{MSE}{J}} = 3.77\sqrt{\dfrac{SSE}{60}} = .4867\sqrt{SSE}$. Thus we wish

$\dfrac{1680}{SSE} > 3.89$ (significance of f) and $.4867\sqrt{SSE} > 10$ ($= 20 - 10$, the difference between the extreme $\bar{x}_{i\bullet}$'s - so no significant differences are identified). These become $431.88 > SSE$ and $SSE > 422.16$, so SSE = 425 will work.

21.

a. Grand mean $= 222.167$, MSTr $= 38,015.1333$, MSE $= 1,681.8333$, and $f = 22.6$. The hypotheses are $H_0 : \mu_1 = \ldots = \mu_6$ vs. H_a : at least two μ_i's differ. Reject H_0 if $f \geq F_{.01,5,78}$ (but since there is no table value for $v_2 = 78$, use $f \geq F_{.01,5,60} = 3.34$) With $22.6 \geq 3.34$, we reject H_0. The data indicates there is a dependence on injection regimen.

b. Assume $t_{.005,78} \approx 2.645$

    i)   Confidence interval for $\mu_1 - \frac{1}{5}(\mu_2 + \mu_3 + \mu_4 + \mu_5 + \mu_6)$:

$$\Sigma c_i \bar{x}_i \pm t_{\alpha/2, I(J-1)} \sqrt{\frac{MSE(\Sigma c_i^2)}{J}}$$

$$= -67.4 \pm (2.645) \sqrt{\frac{1,681.8333(1.2)}{14}} = (-99.16, -35.64).$$

    ii)   Confidence interval for $\frac{1}{4}(\mu_2 + \mu_3 + \mu_4 + \mu_5) - \mu_6$:

$$= 61.75 \pm (2.645) \sqrt{\frac{1,681.8333(1.25)}{14}} = (29.34, 94.16)$$

Section 10.3

23. $J_1 = 5$, $J_2 = 4$, $J_3 = 4$, $J_4 = 5$, $\bar{x}_{1\bullet} = 58.28$, $\bar{x}_{2\bullet} = 55.40$, $\bar{x}_{3\bullet} = 50.85$,

$\bar{x}_{4\bullet} = 45.50$, MSE $= 8.89$. With

$$W_{ij} = Q_{.05,4,14} \cdot \sqrt{\frac{MSE}{2}\left(\frac{1}{J_i} + \frac{1}{J_j}\right)} = 4.11\sqrt{\frac{8.89}{2}\left(\frac{1}{J_i} + \frac{1}{J_j}\right)},$$

$\bar{x}_{1\bullet} - \bar{x}_{2\bullet} \pm W_{12} = (2.88) \pm (5.81)$; $\bar{x}_{1\bullet} - \bar{x}_{3\bullet} \pm W_{13} = (7.43) \pm (5.81)$*;

$\bar{x}_{1\bullet} - \bar{x}_{4\bullet} \pm W_{14} = (12.78) \pm (5.48)$*; $\bar{x}_{2\bullet} - \bar{x}_{3\bullet} \pm W_{23} = (4.55) \pm (6.13)$;

$\bar{x}_{2\bullet} - \bar{x}_{4\bullet} \pm W_{24} = (9.90) \pm (5.81)$*; $\bar{x}_{3\bullet} - \bar{x}_{4\bullet} \pm W_{34} = (5.35) \pm (5.81)$;

*Indicates an interval that doesn't include zero, corresponding to $\mu's$ that are judged significantly different.

$$\underline{\hspace{1.2cm} 4 \hspace{1.5cm} 3 \hspace{1.2cm}} \quad 2 \hspace{1cm} 1$$

$$\underline{\hspace{3cm}}$$

This underscoring pattern does not have a very straightforward interpretation.

25.

a. The distributions of the polyunsaturated fat percentages for each of the four regimens must be normal with equal variances.

b. We have all the $\overline{X}_{i.}$'s , and we need the grand mean:

$$\overline{X}_{..} = \frac{8(43.0)+13(42.4)+17(43.1)+14(43.5)}{52} = \frac{2236.9}{52} = 43.017$$

$$SSTr = \sum_i J_i (\overline{x}_{i.} - \overline{x}_{..})^2 = 8(43.0-43.017)^2 + 13(42.4-43.017)^2$$

$$+17(43.1-43.017)^2 + 13(43.5-43.017)^2 = 8.334$$

and $MSTr = \dfrac{8.334}{3} = 2.778$

$$SSTr = \sum (J_i - 1)s^2 = 7(1.5)^2 + 12(1.3)^2 + 16(1.2)^2 + 13(1.2)^2 = 77.79$$

and $MSE = \dfrac{77.79}{48} = 1.621$. Then $f = \dfrac{MSTr}{MSE} = \dfrac{2.778}{1.621} = 1.714$ Since

$1.714 < F_{.10,3,50} = 2.20$, we can say that the p-value is $> .10$. We do not reject the null hypothesis at significance level .10 (or any smaller), so we conclude that the data suggests no difference in the percentages for the different regimens.

27.

a. Let μ_i = true average folacin content for specimens of brand I. The hypotheses to be tested are $H_0 : \mu_1 = \mu_2 = \mu_3 = \mu_4$ vs. H_a :at least two μ_i's differ .

$$\Sigma\Sigma x_{ij}^2 = 1246.88 \text{ and } \frac{x_{\bullet\bullet}^2}{n} = \frac{(168.4)^2}{24} = 1181.61, \text{ so SST} = 65.27.$$

$$\frac{\Sigma x_{i\bullet}^2}{J_i} = \frac{(57.9)^2}{7} + \frac{(37.5)^2}{5} + \frac{(38.1)^2}{6} + \frac{(34.9)^2}{6} = 1205.10, \text{ so}$$

$$SSTr = 1205.10 - 1181.61 = 23.49.$$

Source	Df	SS	MS	F
Treatments	3	23.49	7.83	3.75
Error	20	41.78	2.09	
Total	23	65.27		

With numerator df = 3 and denominator = 20,

$$F_{.05,3,20} = 3.10 < 3.75 < F_{.01,3,20} = 4.94, \text{ so } .01 < p-value < .05, \text{ and}$$

since the p-value < .05, we reject H_o. At least one of the pairs of brands of green tea has different average folacin content.

b. With $\bar{x}_{i\bullet}$ = 8.27, 7.50, 6.35, and 5.82 for I = 1, 2, 3, 4, we calculate the residuals $x_{ij} - \bar{x}_{i\bullet}$ for all observations. A normal probability plot appears below, and indicates that the distribution of residuals could be normal, so the normality assumption is plausible.

Normal Probability Plot for ANOVA Residuals

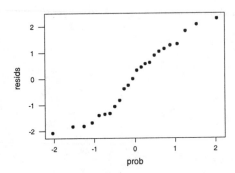

c. $Q_{.05,4,20} = 3.96$ and $W_{ij} = 3.96 \cdot \sqrt{\dfrac{2.09}{2}\left(\dfrac{1}{J_i}+\dfrac{1}{J_j}\right)}$, so the Modified

Tukey intervals are:

Pair	Interval	Pair	Interval
1,2	$.77 \pm 2.37$	2,3	1.15 ± 2.45
1,3	1.92 ± 2.25	2,4	1.68 ± 2.45
1,4	$2.45 \pm 2.25\ *$	3,4	$.53 \pm 2.34$

$$
\begin{array}{cccc}
4 & 3 & 2 & 1
\end{array}
$$

Only Brands 1 and 4 are different from each other.

29.

$$E(SSTr) = E\left(\sum_i J_i \overline{X}_{i\bullet}^2 - n\overline{X}_{\bullet\bullet}^2\right) = \sum_i J_i E\left(\overline{X}_{i\bullet}^2\right) - nE\left(\overline{X}_{\bullet\bullet}^2\right)$$

$$= \sum_i J_i\left[Var\left(\overline{X}_{i\bullet}\right)+\left(E\left(\overline{X}_{i\bullet}\right)\right)^2\right] - n\left[Var\left(\overline{X}_{\bullet\bullet}\right)+\left(E\left(\overline{X}_{\bullet\bullet}\right)\right)^2\right]$$

$$= \sum_i J_i\left[\frac{\sigma^2}{J_i}+\mu_i^2\right] - n\left[\frac{\sigma^2}{n}+\frac{\left(\sum_i J_i\mu_i\right)^2}{n}\right]$$

$$= (I-1)\sigma^2 + \sum_i J_i\left(\mu+\alpha_i\right)^2 - \left[\sum_i J_i\left(\mu+\alpha_i\right)\right]^2$$

$$= (I-1)\sigma^2 + \sum_i J_i\mu^2 + 2\mu\sum_i J_i\alpha_i + \sum_i J_i\alpha_i^2 - \left[\mu\sum_i J_i\right]^2$$

$$= (I-1)\sigma^2 + \sum_i J_i\alpha_i^2 \text{, from which E(MSTr) is obtained through division}$$

by $(I-1)$.

31. With $\sigma = 1$ (any other σ would yield the same Φ), $\alpha_1 = -1$, $\alpha_2 = \alpha_3 = 0$,

$\alpha_4 = 1$, $\Phi^2 = \dfrac{.25\left(5(-1)^2 + 5(0)^2 + 5(0)^2 + 5(1)^2\right)}{1} = 2.5$, $\Phi = 1.58$,

$v_1 = 3$, $v_2 = 14$, and power $\approx .62$.

33. $g(x) = x\left(1 - \dfrac{x}{n}\right) = nu(1-u)$ where $u = \dfrac{x}{n}$, so $h(x) = \int \left[u(1-u)\right]^{-1/2} du$.

From a table of integrals, this gives $h(x) = \arcsin\left(\sqrt{u}\right) = \arcsin\left(\sqrt{\dfrac{x}{n}}\right)$ as the

appropriate transformation.

Supplementary Exercises

35.

a. $H_0 : \mu_1 = \mu_2 = \mu_3 = \mu_4$ vs. H_a : at least two μ_i's differ ; 3.68 is not

$\geq F_{.01,3,20} = 4.94$, thus fail to reject H_0. The means do not appear to differ.

b. We reject H_0 when the p-value < alpha. Since .029 is not < .01, we still fail to reject H_0.

Chapter 10: The Analysis of Variance

37.   Let μ_i = true average amount of motor vibration for each of five bearing brands.
Then the hypotheses are $H_0 : \mu_1 = ... = \mu_5$ vs. H_a :at least two $\mu_i's$ differ.
The ANOVA table follows:

Source	Df	SS	MS	F
Treatments	4	30.855	7.714	8.44
Error	25	22.838	0.914	
Total	29	53.694		

$8.44 > F_{.001,4,25} = 6.49$, so p-value < .001, which is also < .05, so we reject H_0. At
least two of the means differ from one another. The Tukey multiple comparison is
appropriate. $Q_{.05,5,25} = 4.15$ (from Minitab output. Using Table A.10, approximate
with $Q_{.05,5,24} = 4.17$). $W_{ij} = 4.15\sqrt{.914/6} = 1.620$.

Pair	$\bar{x}_{i\bullet} - \bar{x}_{j\bullet}$	Pair	$\bar{x}_{i\bullet} - \bar{x}_{j\bullet}$
1,2	-2.267*	2,4	1.217
1,3	0.016	2,5	2.867*
1,4	-1.050	3,4	-1.066
1,5	0.600	3,5	0.584
2,3	2.283*	4,5	1.650*

*Indicates significant pairs.

5	3	1	4	2

39. $\hat{\theta} = 2.58 - \dfrac{2.63 + 2.13 + 2.41 + 2.49}{4} = .165$, $t_{.025,25} = 2.060$, MSE = .108,
and $\Sigma c_i^2 = (1)^2 + (-.25)^2 + (-.25)^2 + (-.25)^2 + (-.25)^2 = 1.25$, so a 95%
confidence interval for θ is
$.165 \pm 2.060\sqrt{\dfrac{(.108)(1.25)}{6}} = .165 \pm .309 = (-.144,.474)$. This interval does
include zero, so 0 is a plausible value for θ.

41. This is a random effects situation. $H_0 : \sigma_A^2 = 0$ states that variation in laboratories doesn't contribute to variation in percentage. H_o will be rejected in favor of H_a if $f \geq F_{.05,3,8} = 4.07$. SST = 86,078.9897 − 86,077.2224 = 1.7673, SSTr = 1.0559, and SSE = .7114. Thus $f = \dfrac{1.0559/3}{.7114/8} = 3.96$, which is not ≥ 4.07, so H_o cannot be rejected at level .05. Variation in laboratories does not appear to be present.

43. $\sqrt{(I-1)(MSE)(F_{.05,I-1,n-I})} = \sqrt{(2)(2.39)(3.63)} = 4.166$. For $\mu_1 - \mu_2$, $c_1 = 1$, $c_2 = -1$, and $c_3 = 0$, so $\sqrt{\sum \dfrac{c_i^2}{J_i}} = \sqrt{\dfrac{1}{8} + \dfrac{1}{5}} = .570$. Similarly, for $\mu_1 - \mu_3$,

$\sqrt{\sum \dfrac{c_i^2}{J_i}} = \sqrt{\dfrac{1}{8} + \dfrac{1}{6}} = .540$; for $\mu_2 - \mu_3$, $\sqrt{\sum \dfrac{c_i^2}{J_i}} = \sqrt{\dfrac{1}{5} + \dfrac{1}{6}} = .606$, and for

$.5\mu_2 + .5\mu_2 - \mu_3$, $\sqrt{\sum \dfrac{c_i^2}{J_i}} = \sqrt{\dfrac{.5^2}{8} + \dfrac{.5^2}{5} + \dfrac{(-1)^2}{6}} = .498$.

Contrast	Estimate	Interval
$\mu_1 - \mu_2$	25.59 − 26.92 = -1.33	$(-1.33) \pm (.570)(4.166) = (-3.70, 1.04)$
$\mu_1 - \mu_3$	25.59 − 28.17 = -2.58	$(-2.58) \pm (.540)(4.166) = (-4.83, -.33)$
$\mu_2 - \mu_3$	26.92 − 28.17 = -1.25	$(-1.25) \pm (.606)(4.166) = (-3.77, 1.27)$
$.5\mu_2 + .5\mu_2 - \mu_3$	-1.92	$(-1.92) \pm (.498)(4.166) = (-3.99, 0.15)$

45. $Y_{ij} - \overline{Y}_{\bullet\bullet} = c(X_{ij} - \overline{X}_{\bullet\bullet})$ and $\overline{Y}_{i\bullet} - \overline{Y}_{\bullet\bullet} = c(\overline{X}_{i\bullet} - \overline{X}_{\bullet\bullet})$, so each sum of squares involving Y will be the corresponding sum of squares involving X multiplied by c^2. Since F is a ratio of two sums of squares, c^2 appears in both the numerator and denominator so cancels, and F computed from Y_{ij}'s = F computed from X_{ij}'s.

CHAPTER 11

Section 11.1

1.

a. $MSA = \dfrac{30.6}{4} = 7.65$, $MSE = \dfrac{59.2}{12} = 4.93$, $f_A = \dfrac{7.65}{4.93} = 1.55$. Since

1.55 is not $\geq F_{.05,4,12} = 3.26$, don't reject H_{oA}. There is no difference in true average tire lifetime due to different makes of cars.

b. $MSB = \dfrac{44.1}{3} = 14.70$, $f_B = \dfrac{14.70}{4.93} = 2.98$. Since 2.98 is not

$\geq F_{.05,3,12} = 3.49$, don't reject H_{oB}. There is no difference in true average tire lifetime due to different brands of tires.

3. $x_{1\bullet} = 927$, $x_{2\bullet} = 1301$, $x_{3\bullet} = 1764$, $x_{4\bullet} = 2453$, $x_{\bullet 1} = 1347$, $x_{\bullet 2} = 1529$,

$x_{\bullet 3} = 1677$, $x_{\bullet 4} = 1892$, $x_{\bullet\bullet} = 6445$, $\Sigma\Sigma x_{ij}^2 = 2{,}969{,}375$,

$CF = \dfrac{(6445)^2}{16} = 2{,}596{,}126.56$, $SSA = 324{,}082.2$, $SSB = 39{,}934.2$,

$SST = 373{,}248.4$, $SSE = 9232.0$

a.

Source	Df	SS	MS	F
A	3	324,082.2	108,027.4	105.3
B	3	39,934.2	13,311.4	13.0
Error	9	9232.0	1025.8	
Total	15	373,248.4		

Since $F_{.01,3,9} = 6.99$, both H_{oA} and H_{oB} are rejected.

Chapter 11: Multifactor Analysis of Variance

b. $Q_{.01,4,9} = 5.96$, $w = 5.96\sqrt{\dfrac{1025.8}{4}} = 95.4$

i:	1	2	3	4
$\bar{x}_{i\bullet}$:	231.75	325.25	441.00	613.25

All levels of Factor A (gas rate) differ significantly except for 1 and 2

c. $w = 95.4$, as in b

i:	1	2	3	4
$\bar{x}_{\bullet j}$:	336.75	382.25	419.25	473

Only levels 1 and 4 appear to differ significantly.

5.

Source	Df	SS	MS	f
Angle	3	58.16	19.3867	2.5565
Connector	4	246.97	61.7425	8.1419
Error	12	91.00	7.5833	
Total	19	396.13		

$H_0 : \alpha_1 = \alpha_2 = \alpha_3 = \alpha_4 = 0$; H_a :at least one α_i is not zero.
$f_A = 2.5565 < F_{.01,3,12} = 5.95$, so fail to reject H_o. The data fails to indicate any effect due to the angle of pull.

7.

a. CF = 140,454, SST = 3476,

$$SSTr = \frac{(905)^2 + (913)^2 + (936)^2}{18} - 140,454 = 28.78,$$

$$SSBl = \frac{430,295}{3} - 140,454 = 2977.67, \ SSE = 469.55, \ MSTr = 14.39,$$

MSE = 13.81, $f_{Tr} = 1.04$, which is clearly insignificant when compared to $F_{.05,2,51}$.

b. $f_{Bl} = 12.68$, which is significant, and suggests substantial variation among subjects. If we had not controlled for such variation, it might have affected the analysis and conclusions.

9.

Source	Df	SS	MS	f
Treatment	3	81.1944	27.0648	22.36
Block	8	66.5000	8.3125	6.87
Error	24	29.0556	1.2106	
Total	35	176.7500		

$F_{.05,3,24} = 3.01$. Reject H_o. There is an effect due to treatments.

$$Q_{.05,4,24} = 3.90; \; w = (3.90)\sqrt{\frac{1.2106}{9}} = 1.43$$

1	4	3	2
8.56	9.22	10.78	12.44

11. The residual, percentile pairs are (-0.1225, -1.73), (-0.0992, -1.15), (-0.0825, -0.81), (-0.0758, -0.55), (-0.0750, -0.32), (0.0117, -0.10), (0.0283, 0.10), (0.0350, 0.32), (0.0642, 0.55), (0.0708, 0.81), (0.0875, 1.15), (0.1575, 1.73).

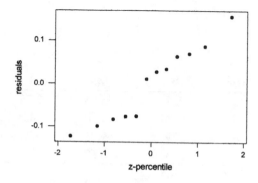

Normal Probability Plot

The pattern is sufficiently linear, so normality is plausible.

13.

a. With $Y_{ij} = X_{ij} + d$, $\bar{Y}_{i\bullet} = \bar{X}_{i\bullet} + d$, $\bar{Y}_{\bullet j} = \bar{X}_{\bullet j} + d$, $\bar{Y}_{\bullet\bullet} = \bar{X}_{\bullet\bullet} + d$, so all
quantities inside the parentheses in (11.5) remain unchanged when the Y
quantities are substituted for the corresponding X's (e.g.,
$\bar{Y}_{i\bullet} - \bar{Y}_{\bullet\bullet} = \bar{X}_{i\bullet} - \bar{X}_{\bullet\bullet}$, etc.).

b. With $Y_{ij} = cX_{ij}$, each sum of squares for Y is the corresponding SS for X
multiplied by c^2. However, when F ratios are formed the c^2 factors cancel, so all
F ratios computed from Y are identical to those computed from X. If
$Y_{ij} = cX_{ij} + d$, the conclusions reached from using the Y's will be identical to
those reached using the X's.

15.

a. $\Sigma\alpha_i^2 = 24$, so $\Phi^2 = \left(\dfrac{3}{4}\right)\left(\dfrac{24}{16}\right) = 1.125$, $\Phi = 1.06$, $\nu_1 = 3$, $\nu_2 = 6$,

and from figure 10.5, power $\approx .2$. For the second alternative, $\Phi = 1.59$, and
power $\approx .43$.

b. $\Phi^2 = \left(\dfrac{1}{J}\right)\Sigma\dfrac{\beta_j^2}{\sigma^2} = \left(\dfrac{4}{5}\right)\left(\dfrac{20}{16}\right) = 1.00$, so $\Phi = 1.00$, $\nu_1 = 4$, $\nu_2 = 12$,

and power $\approx .3$.

Section 11.2

17.

a.

Source	Df	SS	MS	f	$F_{.05}$
Sand	2	705	352.5	3.76	4.26
Fiber	2	1,278	639.0	6.82*	4.26
Sand&Fiber	4	279	69.75	0.74	3.63
Error	9	843	93.67		
Total	17	3,105			

There appears to be an effect due to carbon fiber addition.

Chapter 11: Multifactor Analysis of Variance

b.

Source	Df	SS	MS	f	$F_{.05}$
Sand	2	106.78	53.39	6.54*	4.26
Fiber	2	87.11	43.56	5.33*	4.26
Sand&Fiber	4	8.89	2.22	.27	3.63
Error	9	73.50	8.17		
Total	17	276.28			

There appears to be an effect due to both sand and carbon fiber addition to casting hardness.

c.

Sand%	0	15	30	0	15	30	0	15	30
Fiber%	0	0	0	0.25	0.25	0.25	0.5	0.5	0.5
$\bar{x}$	62	68	69.5	69	71.5	73	68	71.5	74

The plot below indicates some effect due to sand and fiber addition with no significant interaction. This agrees with the statistical analysis in part **b**

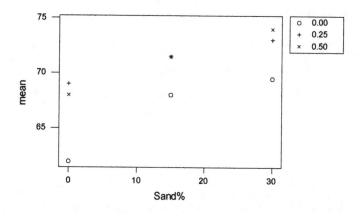

209

19.

a.

	$x_{ij\bullet}$	j = 1	2	3	$x_{i\bullet\bullet}$
	1	16.44	17.27	16.10	49.81
i	2	16.24	17.00	15.91	49.15
	3	16.80	17.37	16.20	50.37
	$x_{\bullet j \bullet}$	49.48	51.64	48.21	$x_{\bullet\bullet\bullet} = 149.33$

CF = 1238.8583

Thus SST = 1240.1525 – 1238.8583 = 1.2942,

$$SSE = 1240.1525 - \frac{2479.9991}{2} = .1530,$$

$$SSA = \frac{(49.81)^2 + (49.15)^2 + (50.37)^2}{6} - 1238.8583 = .1243,$$

$$SSB = 1.0024$$

Source	Df	SS	MS	f	$F_{.01}$
A	2	.1243	.0622	3.66	8.02
B	2	1.0024	.5012	29.48*	8.02
AB	4	.0145	.0036	.21	6.42
Error	9	.1530	.0170		
Total	17	1.2942			

H_{oAB} cannot be rejected, so no significant interaction; H_{oA} cannot be rejected, so varying levels of NaOH does not have a significant impact on total acidity; H_{oB} is rejected: type of coal does appear to affect total acidity.

b. $Q_{.01,3,9} = 5.43$, $w = 5.43\sqrt{\dfrac{.0170}{6}} = .289$

j:	3	1	2
$\bar{x}_{\bullet j \bullet}$	8.035	8.247	8.607

Coal 2 is judged significantly different from both 1 and 3, but these latter two don't differ significantly from each other.

21.

a. $SST = 12,280,103 - \dfrac{(19,143)^2}{30} = 64,954.70$,

$SSE = 12,280,103 - \dfrac{(24,529,699)}{2} = 15,253.50$,

$SSA = \dfrac{122,380,901}{10} - \dfrac{(19,143)^2}{30} = 22,941.80$, $SSB = 22,765.53$,

$SSAB = 64,954.70 - [22,941.80 + 22,765.53 + 15,253.50] = 3993.87$

Source	Df	SS	MS	f
A	2	22,941.80	11,470.90	$\frac{11,470.90}{499.23} = 22.98$
B	4	22,765.53	5691.38	$\frac{5691.38}{499.23} = 11.40$
AB	8	3993.87	499.23	.49
Error	15	15,253.50	1016.90	
Total	29	64,954.70		

b. $f_{AB} = .49$ is clearly not significant. Since $22.98 \geq F_{.05,2,8} = 4.46$, H_{oA} is rejected; since $11.40 \geq F_{.05,4,8} = 3.84$, H_{oB} is also rejected. We conclude that the different cement factors affect flexural strength differently and that batch variability contributes to variation in flexural strength.

23. Summary quantities include $x_{1\bullet\bullet} = 9410$, $x_{2\bullet\bullet} = 8835$, $x_{3\bullet\bullet} = 9234$,

$x_{\bullet 1\bullet} = 5432$, $x_{\bullet 2\bullet} = 5684$, $x_{\bullet 3\bullet} = 5619$,

$x_{\bullet 4\bullet} = 5567$, $x_{\bullet 3\bullet} = 5177$, $x_{\bullet\bullet\bullet} = 27,479$, $CF = 16,779,898.69$,

$\Sigma x_{i\bullet\bullet}^2 = 251,872,081$, $\Sigma x_{\bullet j\bullet}^2 = 151,180,459$, resulting in the accompanying ANOVA table.

Source	Df	SS	MS	f
A	2	11,573.38	5786.69	$\frac{MSA}{MSAB} = 26.70$
B	4	17,930.09	4482.52	$\frac{MSB}{MSAB} = 20.68$
AB	8	1734.17	216.77	$\frac{MSAB}{MSE} = 1.38$
Error	30	4716.67	157.22	
Total	44	35,954.31		

Since $1.38 < F_{.01,8,30} = 3.17$, H_{oG} cannot be rejected, and we continue:

$26.70 \geq F_{.01,2,8} = 8.65$, and $20.68 \geq F_{.01,4,8} = 7.01$, so both H_{oA} and H_{oB} are rejected. Both capping material and the different batches affect compressive strength of concrete cylinders.

25. With $\theta = \alpha_i - \alpha_i'$, $\hat{\theta} = \overline{X}_{i\bullet\bullet} - \overline{X}_{i'\bullet\bullet} = \dfrac{1}{JK}\underset{j}{\Sigma}\underset{k}{\Sigma}\left(X_{ijk} - X_{i'jk}\right)$, and since $i \neq i'$,

$X_{ijk}\, and X_{i'jk}$ are independent for every j, k. Thus

$$Var\!\left(\hat{\theta}\right) = Var\!\left(\overline{X}_{i\bullet\bullet}\right) + Var\!\left(\overline{X}_{i'\bullet\bullet}\right) = \frac{\sigma^2}{JK} + \frac{\sigma^2}{JK} = \frac{2\sigma^2}{JK}\ \text{(because}$$

$Var\!\left(\overline{X}_{i\bullet\bullet}\right) = Var\!\left(\overline{\varepsilon}_{i\bullet\bullet}\right)$ and $Var\!\left(\varepsilon_{ijk}\right) = \sigma^2$) so $\hat{\sigma}_{\hat{\theta}} = \sqrt{\dfrac{2MSE}{JK}}$. The appropriate

number of d.f. is IJ(K – 1), so the C.I. is $\left(\overline{x}_{i\bullet\bullet} - \overline{x}_{i'\bullet\bullet}\right) \pm t_{\alpha/2,IJ(K-1)}\sqrt{\dfrac{2MSE}{JK}}$. For the

data of exercise 19, $\overline{x}_{2\bullet\bullet} = 49.15$, $\overline{x}_{3\bullet\bullet} = 50.37$, MSE = .0170, $t_{.025,9} = 2.262$, J = 3, K = 2, so the C.I. is

$$\left(49.15 - 50.37\right) \pm 2.262\sqrt{\frac{.0370}{6}} = -1.22 \pm .17 = \left(-1.39, -1.05\right).$$

Section 11.3

27.

a.

Source	Df	SS	MS	f	$F_{.05}$
A	2	14,144.44	7072.22	61.06	3.35
B	2	5,511.27	2755.64	23.79	3.35
C	2	244,696.39	122.348.20	1056.24	3.35
AB	4	1,069.62	267.41	2.31	2.73
AC	4	62.67	15.67	.14	2.73
BC	4	331.67	82.92	.72	2.73
ABC	8	1,080.77	135.10	1.17	2.31
Error	27	3,127.50	115.83		
Total	53	270,024.33			

b. The computed f-statistics for all four interaction terms are less than the tabled values for statistical significance at the level .05. This indicates that none of the interactions are statistically significant.

c. The computed f-statistics for all three main effects exceed the tabled value for significance at level .05. All three main effects are statistically significant.

d. $Q_{.05,3,27}$ is not tabled, use $Q_{.05,3,24} = 3.53$, $w = 3.53\sqrt{\dfrac{115.83}{(3)(3)(2)}} = 8.95$. All three levels differ significantly from each other.

29. $I = 3, J = 2, K = 4, L = 4;\ SSA = JKL\sum\left(\bar{x}_{i...} - \bar{x}_{....}\right)^2\ ;$

$SSB = IKL\sum\left(\bar{x}_{.j..} - \bar{x}_{....}\right)^2\ ;\ SSC = IJL\sum\left(\bar{x}_{..k.} - \bar{x}_{....}\right)^2\ .$

For level A: $\bar{x}_{1...} = 3.781$ $\bar{x}_{2...} = 3.625$ $\bar{x}_{3...} = 4.469$

For level B: $\bar{x}_{.1..} = 4.979$ $\bar{x}_{.2..} = 2.938$

For level C: $\bar{x}_{..1.} = 3.417$ $\bar{x}_{..2.} = 5.875$ $\bar{x}_{..3.} = .875$

$\bar{x}_{..4.} = 5.667$

$\bar{x}_{....} = 3.958$

SSA = 12.907;   SSB = 99.976;   SSC = 393.436

a.

Source	Df	SS	MS	f	F.05*
A	2	12.907	6.454	1.04	3.15
B	1	99.976	99.976	16.09	4.00
C	3	393.436	131.145	21.10	2.76
AB	2	1.646	.823	.13	3.15
AC	6	71.021	11.837	1.90	2.25
BC	3	1.542	.514	.08	2.76
ABC	6	9.805	1.634	.26	2.25
Error	72	447.500	6.215		
Total	95	1,037.833			

*use 60 df for denominator of tabled F.

b.   No interaction effects are significant at level .05

c.   Factor B and C main effects are significant at the level .05

d. $Q_{.05,4,72}$ is not tabled, use $Q_{.05,4,60} = 3.74$, $w = 3.74\sqrt{\dfrac{6.215}{(3)(2)(4)}} = 1.90$.

Machine:	3	1	4	2
Mean:	.875	3.417	5.667	5.875

31.

$x_{ij.}$	B_1	B_2	B_3
A_1	210.2	224.9	218.1
A_2	224.1	229.5	221.5
A_3	217.7	230.0	202.0
$x_{.j.}$	652.0	684.4	641.6

$x_{i.k}$	A_1	A_2	A_3
C_1	213.8	222.0	205.0
C_2	225.6	226.5	223.5
C_3	213.8	226.6	221.2
$x_{i..}$	653.2	675.1	649.7

$x_{.jk}$	C_1	C_2	C_3
B_1	213.5	220.5	218.0
B_2	214.3	246.1	224.0
B_3	213.0	209.0	219.6
$x_{..k}$	640.8	675.6	661.6

$\Sigma\Sigma x_{ij.}^2 = 435,382.26 \quad \Sigma\Sigma x_{i.k}^2 = 435,156.74 \quad \Sigma\Sigma x_{.jk}^2 = 435,666.36$

$\Sigma x_{.j.}^2 = 1,305,157.92 \quad \Sigma x_{i..}^2 = 1,304,540.34 \quad \Sigma x_{..k}^2 = 1,304,774.56$

Also, $\Sigma\Sigma\Sigma x_{ijk}^2 = 145,386.40$, $x_{...} = 1978$, CF = 144,906.81, from which we

obtain the ANOVA table displayed in the problem statement. $F_{.01,4,8} = 7.01$, so the

AB and BC interactions are significant (as can b seen from the p-values) and tests for
main effects are not appropriate.

33.

Source	Df	SS	MS	f
A	6	67.32	11.02	
B	6	51.06	8.51	
C	6	5.43	.91	.61
Error	30	44.26	1.48	
Total	48	168.07		

Since $.61 < F_{.05,6,30} = 2.42$, treatment was not effective.

35.

	1	2	3	4	5	
$x_{i..}$	40.68	30.04	44.02	32.14	33.21	$\Sigma x_{i..}^2 = 6630.91$
$x_{.j.}$	29.19	31.61	37.31	40.16	41.82	$\Sigma x_{.j.}^2 = 6605.02$
$x_{..k}$	36.59	36.67	36.03	34.50	36.30	$\Sigma x_{..k}^2 = 6489.92$

$$x_{...} = 180.09, \ CF = 1297.30, \ \Sigma\Sigma x_{ij(k)}^2 = 1358.60$$

37.

Source	Df	MS	f	$F_{.01}$*
A	2	2207.329	2259.29	5.39
B	1	47.255	48.37	7.56
C	2	491.783	503.36	5.39
D	1	.044	.05	7.56
AB	2	15.303	15.66	5.39
AC	4	275.446	281.93	4.02
AD	2	.470	.48	5.39
BC	2	2.141	2.19	5.39
BD	1	.273	.28	7.56
CD	2	.247	.25	5.39
ABC	4	3.714	3.80	4.02
ABD	2	4.072	4.17	5.39
ACD	4	.767	.79	4.02
BCD	2	.280	.29	5.39
ABCD	4	.347	.355	4.02
Error	36	.977		
Total	71			

*Because denominator d.f. for 36 is not tabled, use d.f. = 30

SST = (71)(93.621) = 6,647.091. Computing all other sums of squares and adding them up = 6,645.702. Thus SSABCD = 6,647.091 – 6,645.702 = 1.389 and

$$MSABCD = \frac{1.389}{4} = .347 \,.$$

At level .01 the statistically significant main effects are A, B, C. The interaction AB and AC are also statistically significant. No other interactions are statistically significant.

Section 11.4

39.

Condition	Total	1	2	Contrast	$SS = \frac{(contrast)^2}{24}$
111	315	927	2478	5485	
211	612	1551	3007	1307	A = 71,177.04
121	584	1163	680	1305	B = 70,959.38
221	967	1844	627	199	AB = 1650.04
112	453	297	624	529	C = 11,660.04
212	710	383	681	-53	AC = 117.04
122	737	257	86	57	BC = 135.38
222	1107	370	113	27	ABC = 30.38

a. $\hat{\beta}_1 = \bar{x}_{.2..} - \bar{x}_{....} = \dfrac{584 + 967 + 737 + 1107 - 315 - 612 - 453 - 710}{24} = 54.38$

$\hat{\gamma}_{11}^{AC} = \dfrac{315 - 612 + 584 - 967 - 453 + 710 - 737 + 1107}{24} = 2.21;$

$\hat{\gamma}_{21}^{AC} = -\hat{\gamma}_{11}^{AC} = 2.21.$

b. Factor SS's appear above. With $CF = \dfrac{5485^2}{24} = 1,253,551.04$ and

$\Sigma\Sigma\Sigma\Sigma x_{ijkl}^2 = 1,411,889$, SST = 158,337.96, from which SSE = 2608.7. The ANOVA table appears in the answer section. $F_{.05,1,16} = 4.49$, from which we see that the AB interaction and al the main effects are significant.

41. $\Sigma\Sigma\Sigma\Sigma\Sigma x_{ijklm}^2 = 3,308,143$, $x_{.....} = 11,956$, so

$$CF = \frac{(11,956)^2}{48} = 2,979,535.02\text{, and SST} = 328,607.98.\text{ Each SS is}$$

$$\frac{(effectcontrast)^2}{48}$$ and SSE is obtained by subtraction. The ANOVA table appears

in the answer section. $F_{.05,1,32} \approx 4.15$, a value exceeded by the F ratios for AB
interaction and the four main effects.

43.

Condition/ Effect	$SS = \frac{(contrast)^2}{16}$	f	Condition/ Effect	$SS = \frac{(contrast)^2}{16}$	f
(1)	--		D	414.123	1067.33
A	.436	1.12	AD	.017	< 1
B	.099	< 1	BD	.456	< 1
AB	.497	1.28	ABD	.055	--
C	.109	< 1	CD	2.190	5.64
AC	.078	< 1	ACD	1.020	--
BC	1.404	3.62	BCD	.133	--
ABC	.051	--	ABCD	.681	--

SSE = .051 + .055 + 1.020 + .133 + .681 = 1.940, d.f. = 5, so MSE = .388.
$F_{.05,1,5} = 6.61$, so only the D main effect is significant.

45.

a. The allocation of treatments to blocks is as given in the answer section, with block #1 containing all treatments having an even number of letters in common with both ab and cd, etc.

b. $x_{.....} = 16,898$, so $SST = 9,035,054 - \dfrac{16,898^2}{32} = 111,853.88$. The eight

$block \times replication$ totals are 2091 ($= 618 + 421 + 603 + 449$, the sum of the four observations in block #1 on replication #1), 2092, 2133, 2145, 2113, 2080, 2122, and 2122, so $SSBl = \dfrac{2091^2}{4} + ... + \dfrac{2122^2}{4} - \dfrac{16,898^2}{32} = 898.88$.

The remaining SS's as well as all F ratios appear in the ANOVA table in the answer section. With $F_{.01,1,12} = 9.33$, only the A and B main effects are significant.

47. See the text's answer section.

49.

		A	B	C	D	E	AB	AC	AD	AE	BC	BD	BE	CD	CE	DE
a	70.4	+	-	-	-	-	-	-	-	-	+	+	+	+	+	+
b	72.1	-	+	-	-	-	-	+	+	+	-	-	-	+	+	+
c	70.4	-	-	+	-	-	+	-	+	+	-	+	+	-	-	+
abc	73.8	+	+	+	-	-	+	+	-	-	+	-	-	-	-	+
d	67.4	-	-	-	+	-	+	+	-	+	+	-	+	-	+	-
abd	67.0	+	+	-	+	-	+	-	+	-	-	+	-	-	+	-
acd	66.6	+	-	+	+	-	-	+	+	-	-	-	+	+	-	-
bcd	66.8	-	+	+	+	-	-	-	-	-	+	+	+	-	+	-
e	68.0	-	-	-	-	+	+	+	+	+	-	+	+	-	+	-
abe	67.8	+	+	-	-	+	+	-	-	+	-	-	+	+	-	-
ace	67.5	+	-	+	-	+	-	+	-	+	-	+	-	-	+	-
bce	70.3	-	+	+	-	+	-	-	+	-	+	-	+	-	+	-
ade	64.0	+	-	-	+	+	-	-	+	+	+	-	-	-	-	+
bde	67.9	-	+	-	+	+	-	+	-	-	-	+	+	-	-	+
cde	65.9	-	-	+	+	+	+	-	-	-	-	-	-	+	+	+
abcde	68.0	+	+	+	+	+	+	+	+	+	+	+	+	+	+	+

Thus $SSA = \dfrac{(70.4 - 72.1 - 70.4 + \ldots + 68.0)^2}{16} = 2.250$, SSB = 7.840, SSC = .360, SSD = 52.563, SSE = 10.240, SSAB = 1.563, SSAC = 7.563, SSAD = .090, SSAE = 4.203, SSBC = 2.103, SSBD = .010, SSBE = .123, SSCD = .010, SSCE = .063, SSDE = 4.840, Error SS = sum of two factor SS's = 20.568, Error MS = 2.057, $F_{.01,1,10} = 10.04$, so only the D main effect is significant.

Supplementary Exercises

51.

Source	Df	SS	MS	f
A	1	322.667	322.667	980.38
B	3	35.623	11.874	36.08
AB	3	8.557	2.852	8.67
Error	16	5.266	.329	
Total	23	372.113		

We first test the null hypothesis of no interactions ($H_0 : \gamma_{ij} = 0$ for all I, j). H_o will

be rejected at level .05 if $f_{AB} = \dfrac{MSAB}{MSE} \geq F_{.05,3,16} = 3.24$. Because $8.67 \geq 3.24$,

H_o is rejected. Because we have concluded that interaction is present, tests for main effects are not appropriate.

53. Let A = spray volume, B = belt speed, C = brand.

Condition	Total	1	2	Contrast	$SS = \frac{(contrast)^2}{16}$
(1)	76	129	289	592	21,904.00
A	53	160	303	22	30.25
B	62	143	13	48	144.00
AB	98	160	9	134	1122.25
C	88	-23	31	14	12.25
AC	55	36	17	-4	1.00
BC	59	-33	59	-14	12.25
ABC	101	42	75	16	16.00

The ANOVA table is as follows:

Effect	Df	MS	f
A	1	30.25	6.72
B	1	144.00	32.00
AB	1	1122.25	249.39
C	1	12.25	2.72
AC	1	1.00	.22
BC	1	12.25	2.72
ABC	1	16.00	3.56
Error	8	4.50	
Total	15		

$F_{.05,1,8} = 5.32$, so all of the main effects are significant at level .05, but none of the interactions are significant.

55.

a.

Effect	%Iron	1	2	3	Effect Contrast	SS
	7	18	37	174	684	
A	11	19	137	510	144	1296
B	7	62	169	50	36	81
AB	12	75	341	94	0	0
C	21	79	9	14	272	4624
AC	41	90	41	22	32	64
BC	27	165	47	2	12	9
ABC	48	176	47	-2	-4	1
D	28	4	1	100	336	7056
AD	51	5	13	172	44	121
BD	33	20	11	32	8	4
ABD	57	21	11	0	0	0
CD	70	23	1	12	72	324
ACD	95	24	1	0	-32	64
BCD	77	25	1	0	-12	9
ABCD	99	22	-3	-4	-4	1

We use $estimate = \dfrac{contrast}{2^p}$ when n = 1 (see p 472 of text) to get

$$\hat{\alpha}_1 = \frac{144}{2^4} = \frac{144}{16} = 9.00,\ \hat{\beta}_1 = \frac{36}{16} = 2.25,\ \hat{\delta}_1 = \frac{272}{16} = 17.00,$$

$$\hat{\gamma}_1 = \frac{336}{16} = 21.00.\ \text{Similarly,}\ \left(\widehat{\alpha\beta}\right)_{11} = 0,\ \left(\widehat{\alpha\delta}\right)_{11} = 2.00,$$

$$\left(\widehat{\alpha\gamma}\right)_{11} = 2.75,\ \left(\widehat{\beta\delta}\right)_{11} = .75,\ \left(\widehat{\beta\gamma}\right)_{11} = .50,\ \text{and}\ \left(\widehat{\delta\gamma}\right)_{11} = 4.50.$$

b.

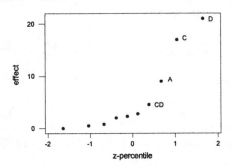

The plot suggests main effects A, C, and D are quite important, and perhaps the interaction CD as well. (See answer section for comment.)

57. The ANOVA table is:

Source	df	SS	MS	f	$F_{.01}$
A	2	34,436	17,218	436.92	5.49
B	2	105,793	52,897	1342.30	5.49
C	2	516,398	258,199	6552.04	5.49
AB	4	6,868	1,717	43.57	4.11
AC	4	10,922	2,731	69.29	4.11
BC	4	10,178	2,545	64.57	4.11
ABC	8	6,713	839	21.30	3.26
Error	27	1,064	39		
Total	53	692,372			

All calculated f values are greater than their respective tabled values, so all effects, including the interaction effects, are significant at level .01.

59. Based on the p-values in the ANOVA table, statistically significant factors at the level .01 are adhesive type and cure time. The conductor material does not have a statistically significant effect on bond strength. There are no significant interactions.

61. $SSA = \sum_i \sum_j (\overline{X}_{i...} - \overline{X}_{....})^2 = \frac{1}{N}\Sigma X_{i...}^2 - \frac{X_{....}^2}{N}$, with similar expressions for

SSB, SSC, and SSD, each having $N-1$ df.

$SST = \sum_i \sum_j (X_{ij(kl)} - \overline{X}_{....})^2 = \sum_i \sum_j X_{ij(kl)}^2 - \frac{X_{....}^2}{N}$ with N^2-1 df, leaving

$N^2 - 1 - 4(N-1)$ df for error.

	1	2	3	4	5	Σx^2
$x_{i...}$:	482	446	464	468	434	1,053,916
$x_{.j..}$:	470	451	440	482	451	1,053,626
$x_{..k.}$:	372	429	484	528	481	1,066,826
$x_{...l}$:	340	417	466	537	534	1,080,170

Also, $\Sigma\Sigma x_{ij(kl)}^2 = 220{,}378$, $x_{....} = 2294$, and CF = 210,497.44

Source	df	SS	MS	f	$F_{.05}$
A	4	285.76	71.44	.594	3.84
B	4	227.76	56.94	.473	3.84
C	4	2867.76	716.94	5.958*	3.84
D	4	5536.56	1384.14	11.502*	3.84
Error	8	962.72	120.34		
Total	24				

H_{oA} and H_{oB} cannot be rejected, while while H_{oC} and H_{oD} are rejected.

CHAPTER 12

Section 12.1

1.

    a. Stem and Leaf display of temp:

$$
\begin{array}{r|l}
17 & 0 \\
17 & 23 \\
17 & 445 \\
17 & 67 \\
17 & \\
18 & 0000011 \\
18 & 2222 \\
18 & 445 \\
18 & 6 \\
18 & 8 \\
\end{array}
$$

    stem = tens

    leaf = ones

180 appears to be a typical value for this data. The distribution is reasonably symmetric in appearance and somewhat bell-shaped. The variation in the data is fairly small since the range of values ($188 - 170 = 18$) is fairly small compared to the typical value of 180.

$$
\begin{array}{r|l}
0 & 889 \\
1 & 0000 \\
1 & 3 \\
1 & 4444 \\
1 & 66 \\
1 & 8889 \\
2 & 11 \\
2 & \\
2 & 5 \\
2 & 6 \\
2 & \\
3 & 00 \\
\end{array}
$$

    stem = ones

    leaf = tenths

For the ratio data, a typical value is around 1.6 and the distribution appears to be positively skewed. The variation in the data is large since the range of the data $(3.08 - .84 = 2.24)$ is very large compared to the typical value of 1.6. The two largest values could be outliers.

b. The efficiency ratio is not uniquely determined by temperature since there are several instances in the data of equal temperatures associated with different efficiency ratios. For example, the five observations with temperatures of 180 each have different efficiency ratios.

c. A scatter plot of the data appears below. The points exhibit quite a bit of variation and do not appear to fall close to any straight line or simple curve.

3. A scatter plot of the data appears below. The points fall very close to a straight line with an intercept of approximately 0 and a slope of about 1. This suggests that the two methods are producing substantially the same concentration measurements.

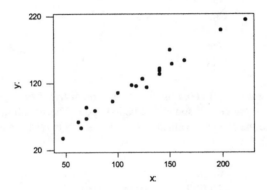

Chapter 12: Simple Linear Regression and Correlation

5.

    a. The scatter plot with axes intersecting at (0,0) is shown below.

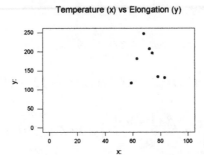

    b. The scatter plot with axes intersecting at (55, 100) is shown below.

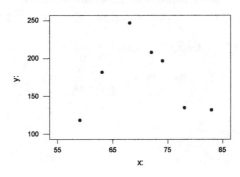

    c. A parabola appears to provide a good fit to both graphs.

7.

    a. $\mu_{Y \cdot 2500} = 1800 + 1.3(2500) = 5050$

    b. expected change = slope = $\beta_1 = 1.3$

    c. expected change = $100\beta_1 = 130$

    d. expected change = $-100\beta_1 = -130$

9.

a. β_1 = expected change in flow rate (y) associated with a one inch increase in pressure drop (x) = .095.

b. We expect flow rate to decrease by $5\beta_1 = .475$.

c. $\mu_{Y\cdot10} = -.12 + .095(10) = .83$, and $\mu_{Y\cdot15} = -.12 + .095(15) = 1.305$.

d. $P(Y > .835) = P\left(Z > \dfrac{.835 - .830}{.025}\right) = P(Z > .20) = .4207$

   $P(Y > .840) = P\left(Z > \dfrac{.840 - .830}{.025}\right) = P(Z > .40) = .3446$

e. Let Y_1 and Y_2 denote pressure drops for flow rates of 10 and 11, respectively. Then $\mu_{Y\cdot11} = .925$, so $Y_1 - Y_2$ has expected value .830 - .925 = -.095, and s.d.

   $\sqrt{(.025)^2 + (.025)^2} = .035355$. Thus

   $P(Y_1 > Y_2) = P(Y_1 - Y_2 > 0) = P\left(z > \dfrac{+.095}{.035355}\right) = P(Z > 2.69) = .0036$

11.

a. β_1 = expected change for a one degree increase = -.01, and $10\beta_1 = -.1$ is the expected change for a 10 degree increase.

b. $\mu_{Y\cdot200} = 5.00 - .01(200) = 3$, and $\mu_{Y\cdot250} = 2.5$.

c. The probability that the first observation is between 2.4 and 2.6 is

$$P(2.4 \le Y \le 2.6) = P\left(\frac{2.4 - 2.5}{.075} \le Z \le \frac{2.6 - 2.5}{.075}\right)$$

$= P(-1.33 \le Z \le 1.33) = .8164$. The probability that any particular one of the other four observations is between 2.4 and 2.6 is also .8164, so the probability that all five are between 2.4 and 2.6 is $(.8164)^5 = .3627$.

d. Let Y_1 and Y_2 denote the times at the higher and lower temperatures, respectively. Then $Y_1 - Y_2$ has expected value

$5.00 - .01(x + 1) - (5.00 - .01x) = -.01$. The standard deviation of $Y_1 - Y_2$

is $\sqrt{(.075)^2 + (.075)^2} = .10607$. Thus

$$P(Y_1 - Y_2 > 0) = P\left(z > \frac{-(-.01)}{.10607}\right) = P(Z > .09) = .4641.$$

Section 12.2

13. For this data, n = 4, $\Sigma x_i = 200$, $\Sigma y_i = 5.37$, $\Sigma x_i^2 = 12.000$, $\Sigma y_i^2 = 9.3501$,

$\Sigma x_i y_i = 333$. $S_{xx} = 12{,}000 - \dfrac{(200)^2}{4} = 2000$,

$S_{yy} = 9.3501 - \dfrac{(5.37)^2}{4} = 2.140875$, and $S_{xy} = 333 - \dfrac{(200)(5.37)}{4} = 64.5$.

$\hat{\beta}_1 = \dfrac{S_{xy}}{S_{xx}} = \dfrac{64.5}{2000} = .03225$ and $\hat{\beta}_0 = \dfrac{5.37}{4} - (.03225)\dfrac{200}{4} = -.27000$.

$SSE = S_{yy} - \hat{\beta}_1 S_{xy} = 2.14085 - (.03225)(64.5) = .060750$.

$r^2 = 1 - \dfrac{SSE}{SST} = 1 - \dfrac{.060750}{2.14085} = .972$. This is a very high value of r^2, which

confirms the authors' claim that there is a strong linear relationship between the two variables.

15.

a. The following stem and leaf display shows that: a typical value for this data is a number in the low 40's. there is some positive skew in the data. There are some potential outliers (79.5 and 80.0), and there is a reasonably large amount of variation in the data (e.g., the spread 80.0-29.8 = 50.2 is large compared with the typical values in the low 40's).

```
2|9
3|33              stem = tens
3|5566677889      leaf = ones
4|1223
4|56689
5|1
5|
6|2
6|9
7|
7|9
8|0
```

b. No, the strength values are not uniquely determined by the MoE values. For example, note that the two pairs of observations having strength values of 42.8 have different MoE values.

c. The least squares line is $\hat{y} = 3.2925 + .10748x$. For a beam whose modulus of elasticity is x = 40, the predicted strength would be $\hat{y} = 3.2925 + .10748(40) = 7.59$. The value x = 100 isfar beyond the range of the x values in the data, so it would be dangerous (i.e., potentially misleading) to extrapolated the linear relationship that far.

d. From the output, SSE = 18.736, SST = 71.605, and the coefficient of determination is $r^2 = .738$ (or 73.8%). The r^2 value is large, which suggests that the linear relationship is a useful approximation to the true relationship between these two variables.

17.   Note: n = 23 in this study.
    a.   For a one (mg/cm^2) increase in dissolved material, one would expect a .144 (g/l) increase in calcium content. Secondly, 86% of the observed variation in calcium content can be attributed to the simple linear regression relationship between calcium content and dissolved material.

    b. $\mu_{y\cdot50} = 3.678 + .144(50) = 10.878$

    c. $r^2 = .86 = 1 - \dfrac{SSE}{SST}$, so

           $SSE = (SST)(1 - .86) = (320.398)(.14) = 44.85572$. Then

           $s = \sqrt{\dfrac{SSE}{n-2}} = \sqrt{\dfrac{44.85572}{21}} = 1.46$

19. $N = 14$, $\Sigma x_i = 3300$, $\Sigma y_i = 5010$, $\Sigma x_i^2 = 913,750$, $\Sigma y_i^2 = 2,207,100$,
       $\Sigma x_i y_i = 1,413,500$

    a. $\hat{\beta}_1 = \dfrac{3,256,000}{1,902,500} = 1.71143233$, $\hat{\beta}_0 = -45.55190543$, so we use the equation $y = -45.5519 + 1.7114x$.

    b. $\hat{\mu}_{Y\cdot225} = -45.5519 + 1.7114(225) = 339.51$

    c.   Estimated expected change $= -50\hat{\beta}_1 = -85.57$

    d.   No, the value 500 is outside the range of x values for which observations were available (the danger of extrapolation).

21.

a. Verification

b. $\hat{\beta}_1 = \dfrac{491.4}{744.16} = .66034186$, $\hat{\beta}_0 = -2.18247148$

c. predicted $y = \hat{\beta}_0 + \hat{\beta}_1(15) = 7.72$

d. $\hat{\mu}_{Y.15} = \hat{\beta}_0 + \hat{\beta}_1(15) = 7.72$

23.

a. Using the $y_i's$ given to one decimal place accuracy is the answer to Exercise 19, $SSE = (150 - 125.6)^2 + ... + (670 - 639.0)^2 = 16,213.64$. The computation formula gives
$SSE = 2,207,100 - (-45.55190543)(5010) - (1.71143233)(1,413,500)$
$= 16,205.45$

b. $SST = 2,207,100 - \dfrac{(5010)^2}{14} = 414,235.71$ so

$r^2 = 1 - \dfrac{16,205.45}{414,235.71} = .961$.

25. Substitution of $\hat{\beta}_0 = \dfrac{\Sigma y_i - \hat{\beta}_1 \Sigma x_i}{n}$ and $\hat{\beta}_1$ for b_0 and b_1 on the left hand side of

the normal equations yields $\dfrac{n(\Sigma y_i - \hat{\beta}_1 \Sigma x_i)}{n} + \hat{\beta}_1 \Sigma x_i = \Sigma y_i$ from the first

equation and $\dfrac{\Sigma x_i(\Sigma y_i - \hat{\beta}_1 \Sigma x_i)}{n} + \hat{\beta}_1 \Sigma x_i^2 = \dfrac{\Sigma x_i \Sigma y_i}{n} + \dfrac{\hat{\beta}_1(n\Sigma x_i^2 - (\Sigma x_i)^2)}{n}$

$\dfrac{\Sigma x_i \Sigma y_i}{n} + \dfrac{n\Sigma x_i y_i}{n} - \dfrac{\Sigma x_i \Sigma y_i}{n} = \Sigma x_i y_i$ from the second equation.

27. We wish to find b_1 to minimize $\Sigma(y_i - b_1 x_i)^2 = f(b_1)$. Equating $f'(b_1)$ to 0

yields $2\Sigma(y_i - b_1 x_i)(-x_i) = 0$ so $\Sigma x_i y_i = b_1 \Sigma x_i^2$ and $b_1 = \dfrac{\Sigma x_i y_i}{\Sigma x_i^2}$. The least

squares estimator of $\hat{\beta}_1$ is thus $\hat{\beta}_1 = \dfrac{\Sigma x_i Y_i}{\Sigma x_i^2}$.

29. For data set #1, $r^2 = .43$ and $\hat{\sigma} = s = 4.03$; whereas these quantities are .99 and 4.03 for #2, and .99 and 1.90 for #3. In general, one hopes for both large r^2 (large % of variation explained) and small s (indicating that observations don't deviate much from the estimated line). Simple linear regression would thus seem to be most effective in the third situation.

Section 12.3

31.

    a. $\hat{\beta}_1 = -.00736023$, $\hat{\beta}_0 = 1.41122185$, so

       $SSE = 7.8518 - (1.41122185)(10.68) - (-.00736023)(987.645) = .04925$,

       $s^2 = .003788$, $s = .06155$.

$$\hat{\sigma}_{\hat{\beta}_1}^2 = \frac{s^2}{\Sigma x_i^2 - (\Sigma x_i)^2 / n} = \frac{.003788}{3662.25} = .00000103,$$

       $\hat{\sigma}_{\hat{\beta}_1} = s_{\hat{\beta}_1} = $ estimated s.d. of $\hat{\beta}_1 = \sqrt{.00000103} = .001017$.

    b. $-.00736 \pm (2.160)(.001017) = -.00736 \pm .00220 = (-.00956, -.00516)$

33.

    a. From the printout in Exercise 15, the error d.f. $= n - 2 = 25$, $t_{.025,25} = 2.060$.

       The confidence interval is then

       $\hat{\beta}_1 \pm t_{.025,25} \cdot s_{\hat{\beta}_1} = .10748 \pm (2.060)(.01280) = (.081, .134)$. Therefore, we estimate

       with a high degree of confidence that the true average change in strength associated with a 1 Gpa increase in modulus of elasticity is between .081 MPa and .134 MPa.

b. We wish to test $H_o : \beta_1 = .1$ vs. $H_a : \beta_1 > .1$. The calculated t statistic is

$$t = \frac{\hat{\beta}_1 - .1}{s_{\hat{\beta}_1}} = \frac{.10748 - .1}{.01280} = .58 \text{, which yields a p-value of .277. A large p-value}$$

such as this would not lead to rejecting H_o, so there is not enough evidence to contradict the prior belief.

35.

a. We want a 95% CI for β_1: $\hat{\beta}_1 \pm t_{.025,15} \cdot s_{\hat{\beta}_1}$. First, we need our point estimate,

$\hat{\beta}_1$. Using the given summary statistics, $S_{xx} = 3056.69 - \dfrac{(222.1)^2}{17} = 155.019$,

$S_{xy} = 2759.6 - \dfrac{(222.1)(193)}{17} = 238.112$, and $\hat{\beta}_1 = \dfrac{S_{xy}}{S_{xx}} = \dfrac{238.112}{115.019} = 1.536$.

We need $\hat{\beta}_0 = \dfrac{193 - (1.536)(222.1)}{17} = -8.715$ to calculate the SSE:

$SSE = 2975 - (-8.715)(193) - (1.536)(2759.6) = 418.2494$. Then

$s = \sqrt{\dfrac{418.2494}{15}} = 5.28$ and $s_{\hat{\beta}_1} = \dfrac{5.28}{\sqrt{155.019}} = .424$. With $t_{.025,15} = 2.131$, our CI

is $1.536 \pm 2.131 \cdot (.424) = (.632, 2.440)$. With 95% confidence, we estimate that the change in reported nausea percentage for every one-unit change in motion sickness dose is between .632 and 2.440.

b. We test the hypotheses $H_o : \beta_1 = 0$ vs $H_a : \beta_1 \neq 0$, and the test statistic is

$t = \dfrac{1.536}{.424} = 3.6226$. With df=15, the two-tailed p-value = 2P(t > 3.6226) = 2(

.001) = .002. With a p-value of .002, we would reject the null hypothesis at most reasonable significance levels. This suggests that there is a useful linear relationship between motion sickness dose and reported nausea.

c. No. A regression model is only useful for estimating values of nausea % when using dosages between 6.0 and 17.6 – the range of values sampled.

d. Removing the point (6.0, 2.50), the new summary stats are: n = 16, , $\Sigma x_i = 216.1$, $\Sigma y_i = 191.5$, $\Sigma x_i^2 = 3020.69$, $\Sigma y_i^2 = 2968.75$, $\Sigma x_i y_i = 2744.6$, and then $\hat{\beta}_1 = 1.561$, $\hat{\beta}_0 = -9.118$, SSE = 430.5264, $s = 5.55$, $s_{\hat{\beta}_1} = .551$, and the new CI is $1.561 \pm 2.145 \cdot (.551)$, or (.379, 2.743). The interval is a little wider. But removing the one observation did not change it that much. The observation does not seem to be exerting undue influence.

37.

a. $n = 10$, $\Sigma x_i = 2615$, $\Sigma y_i = 39.20$, $\Sigma x_i^2 = 860{,}675$, $\Sigma y_i^2 = 161.94$,

$\Sigma x_i y_i = 11{,}453.5$, so $\hat{\beta}_1 = \dfrac{12{,}027}{1{,}768{,}525} = .00680058$,

$\hat{\beta}_0 = 2.14164770$, from which SSE $= .09696713$, $s = .11009492$

$s = .11009492 \doteq .110 = \hat{\sigma}$, $\hat{\sigma}_{\hat{\beta}_1} = \dfrac{.110}{\sqrt{176{,}852}} = .000262$

b. We wish to test $H_o : \beta_1 = .0060$ vs $H_a : \beta_1 \neq .0060$. At level .10, H_o is rejected if either $t \geq t_{.05,8} = 1.860$ or $t \leq -t_{.05,8} = -1.860$. Since

$t = \dfrac{.0068 - .0060}{.000262} = 3.06 \geq 1.1860$, H_o is rejected.

39. SSE $= 124{,}039.58 - (72.958547)(1574.8) - (.04103377)(222657.88) = 7.9679$, and SST $= 39.828$

Source	df	SS	MS	f
Regr	1	31.860	31.860	18.0
Error	18	7.968	1.77	
Total	19	39.828		

Let's use $\alpha = .001$. Then $F_{.001,1,18} = 15.38 < 18.0$, so $H_o : \beta_1 = 0$ is rejected and the model is judged useful. $s = \sqrt{1.77} = 1.33041347$, $S_{xx} = 18{,}921.8295$,

so $t = \dfrac{.04103377}{1.33041347 / \sqrt{18{,}921.8295}} = 4.2426$, and

$t^2 = (4.2426)^2 = 18.0 = f$.

41.

a. Let $c = n\Sigma x_i^2 - (\Sigma x_i)^2$. Then $E(\hat{\beta}_1) = \frac{1}{c} E[n\Sigma x_i Y_i ... Y_i - (\Sigma x_i)..(\Sigma x_i)(\Sigma Y_i)]$

$= \frac{n}{c} \sum x_i E(Y_i) - \frac{\Sigma x_i}{c} \sum E(Y_i) = \frac{n}{c} \sum x_i (\beta_0 + \beta_1 x_i) - \frac{\Sigma x_i}{c} \sum (\beta_0 + \beta_1 x_i)$

$\frac{\beta_1}{c}[n\Sigma x_i^2 - (\Sigma x_i)^2] = \beta_1$.

b. With $c = \Sigma(x_i - \bar{x})^2$, $\hat{\beta}_1 = \frac{1}{c}\Sigma(x_i - \bar{x})(Y_i - \bar{Y}) = \frac{1}{c}\Sigma(x_i - \bar{x})Y_i$ (since

$\Sigma(x_i - \bar{x})\bar{Y} = \bar{Y}\Sigma(x_i - \bar{x}) = \bar{Y} \cdot 0 = 0$), so $V(\hat{\beta}_1) = \frac{1}{c^2}\Sigma(x_i - \bar{x})^2 Var(Y_i)$

$= \frac{1}{c^2}\Sigma(x_i - \bar{x})^2 \cdot \sigma^2 = \frac{\sigma^2}{\Sigma(x_i - \bar{x})^2} = \frac{\sigma^2}{\Sigma x_i^2 - (\Sigma x_i)^2 / n}$, as desired.

43. The numerator of d is $|1 - 2| = 1$, and the denominator is $\dfrac{4\sqrt{14}}{\sqrt{324.40}} = .831$, so

$d = \dfrac{1}{.831} = 1.20$. The approximate power curve is for $n - 2$ df $= 13$, and β is read from Table A.17 as approximately .1.

Section 12.4

45.

a. We wish to find a 90% CI for $\mu_{y \cdot 125}$: $\hat{y}_{125} = 78.088$, $t_{.05,18} = 1.734$, and

$s_{\hat{y}} = s\sqrt{\dfrac{1}{20} + \dfrac{(125 - 140.895)^2}{18,921.8295}} = .3349$. Putting it together, we get

$78.088 \pm 1.734(.3349) = (77.5073, 78.6687)$

b. We want a 90% PI: Only the standard error changes:

$s_{\hat{y}} = s\sqrt{1 + \dfrac{1}{20} + \dfrac{(125 - 140.895)^2}{18,921.8295}} = 1.3719$, so the PI is

$78.088 \pm 1.734(1.3719) = (75.7091, 80.4669)$

c. Because the x^* of 115 is farther away from $\bar{x}$ than the previous value, the term $\left(x^* - \bar{x}\right)^2$ will be larger, making the standard error larger, and thus the width of the interval is wider.

d. We would be testing to see if the filtration rate were 125 kg-DS/m/h, would the average moisture content of the compressed pellets be less than 80%.

47.

a. $\hat{y}_{(40)} = -1.128 + .82697(40) = 31.95$, $t_{.025,13} = 2.160$; a 95% PI for runoff is

$$31.95 \pm 2.160\sqrt{(5.24)^2 + (1.44)^2} = 31.95 \pm 11.74 = (20.21, 43.69).$$

No, the resulting interval is very wide, therefore the available information is not very precise.

b. $\Sigma x = 798, \Sigma x^2 = 63,040$ which gives $S_{xx} = 20,586.4$, which in turn gives

$$s_{\hat{y}_{(50)}} = 5.24\sqrt{\frac{1}{15} + \frac{(50 - 53.20)^2}{20,586.4}} = 1.358,$$ so the PI for runoff when x = 50 is

$$40.22 \pm 2.160\sqrt{(5.24)^2 + (1.358)^2} = 40.22 \pm 11.69 = (28.53, 51.92).$$

The simultaneous prediction level for the two intervals is at least $100(1 - 2\alpha)\% = 90\%$.

49. 95% CI: (462.1, 597.7); midpoint = 529.9; $t_{.025,8} = 2.306$;

$$529.9 + (2.306)\left(\hat{s}_{\hat{\beta}_0 + \hat{\beta}_1(15)}\right) = 597.7$$

$$\hat{s}_{\hat{\beta}_0 + \hat{\beta}_1(15)} = 29.402$$

99% CI: $529.9 \pm (3.355)(29.402) = (431.3, 628.5)$

51.

a. 0.40 is closer to $\bar{x}$.

b. $\hat{\beta}_0 + \hat{\beta}_1(0.40) \pm t_{\alpha/2,n-2} \cdot \left(\hat{s}_{\hat{\beta}_0 + \hat{\beta}_1(0.40)} \right)$ or $0.8104 \pm (2.101)(0.0311)$
$= (0.745, 0.876)$

c. $\hat{\beta}_0 + \hat{\beta}_1(1.20) \pm t_{\alpha/2,n-2} \cdot \sqrt{s^2 + s^2{}_{\hat{\beta}_0 + \hat{\beta}_1(1.20)}}$ or
$0.2912 \pm (2.101) \cdot \sqrt{(0.1049)^2 + (0.0352)^2} = (.059, .523)$

53. Choice **a** will be the smallest, with **d** being largest. **a** is less than **b** and **c** (obviously), and **b** and **c** are both smaller than **d**. Nothing can be said about the relationship between **b** and **c**.

55.

a. $x_2 = x_3 = 12$, yet $y_2 \neq y_3$

b.

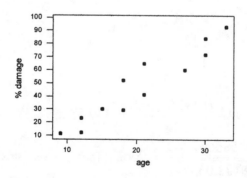

Based on a scatterplot of the data, a simple linear regression model does seem a reasonable way to describe the relationship between the two variables.

c. $\hat{\beta}_1 = \dfrac{2296}{699} = 3.284692$, $\hat{\beta}_0 - 19.669528$, $y = -19.67 + 3.285x$

d. $SSE = 35,634 - (-19.669528)(572) - (3.284692)(14,022) = 827.0188$

, $s^2 = 82.70188, s = 9.094$.

$$s_{\hat{\beta}_0 + \hat{\beta}_1(20)} = 9.094\sqrt{\frac{1}{12} + \frac{12(20 - 20.5)^2}{8388}} = 2.6308,$$

$\hat{\beta}_0 + \hat{\beta}_1(20) = 46.03$, $t_{.025,10} = 2.228$. The PI is

$46.03 \pm 2.228\sqrt{s^2 + s_{\hat{\beta}_0 + \hat{\beta}_1(20)}^2} = 46.03 \pm 21.09 = (24.94, 67.12)$.

Section 12.5

57. Most people acquire a license as soon as they become eligible. If, for example, the minimum age for obtaining a license is 16, then the time since acquiring a license, y, is usually related to age by the equation $y \approx x - 16$, which is the equation of a straight line. In other words, the majority of people in a sample will have y values that closely follow the line $y = x - 16$.

59.

a. $S_{xx} = 251,970 - \frac{(1950)^2}{18} = 40,720$,

$S_{yy} = 130.6074 - \frac{(47.92)^2}{18} = 3.033711$, and

$S_{xy} = 5530.92 - \frac{(1950)(47.92)}{18} = 339.586667$, so

$r = \frac{339.586667}{\sqrt{40,720}\sqrt{3.033711}} = .9662$. There is a very strong positive

correlation between the two variables.

b. Because the association between the variables is positive, the specimen with the larger shear force will tend to have a larger percent dry fiber weight.

c. Changing the units of measurement on either (or both) variables will have no effect on the calculated value of r, because any change in units will affect both the numerator and denominator of r by exactly the same multiplicative constant.

d. $r^2 = (.966)^2 = .933$

e. $H_o : \rho = 0$ vs $H_a : \rho > 0$. $t = \dfrac{r\sqrt{n-2}}{\sqrt{1-r^2}}$; Reject H_o at level .01 if

$t \geq t_{.01,16} = 2.583$. $t = \dfrac{.966\sqrt{16}}{\sqrt{1-.966^2}} = 14.94 \geq 2.583$, so H_o should be

rejected . The data indicates a positive linear relationship between the two variables.

61.

a. We are testing $H_o : \rho = 0$ vs $H_a : \rho > 0$.

$r = \dfrac{7377.704}{\sqrt{36.9839}\sqrt{2,628,930.359}} = .7482$, and

$t = \dfrac{.7482\sqrt{12}}{\sqrt{1-.7482^2}} = 3.9066$. We reject H_o since

$t = 3.9066 \geq t_{.05,12} = 1.782$. There is evidence that a positive correlation exists between maximum lactate level and muscular endurance.

b. We are looking for r^2, the coefficient of determination. $r^2 = (.7482)^2 = .5598$. It is the same no matter which variable is the predictor.

63. $n = 6$, $\Sigma x_i = 111.71$, $\Sigma x_i^2 = 2{,}724.7643$, $\Sigma y_i = 2.9$, $\Sigma y_i^2 = 1.6572$, and $\Sigma x_i y_i = 63.915$.

$$r = \frac{(6)(63.915) - (111.71)(2.9)}{\sqrt{(6)(2{,}724.7943) - (111.73)^2} \cdot \sqrt{(6)(1.6572) - (2.9)^2}} = .7729 .$$

$H_o : \rho_1 = 0$ vs $H_a : \rho \neq 0$; Reject H_o at level .05 if $|t| \geq t_{.025,4} = 2.776$.

$$t = \frac{(.7729)\sqrt{4}}{\sqrt{1 - (.7729)^2}} = 2.436 .$$ Fail to reject H_o. The data does not indicate that the

population correlation coefficient differs from 0. This result may seem surprising due to the relatively large size of r (.77), however, it can be attributed to a small sample size (6).

65.

a. Although the normal probability plot of the x's appears somewhat curved, such a pattern is not terribly unusual when n is small; the test of normality presented in section 14.2 (p. 625) does not reject the hypothesis of population normality. The normal probability plot of the y's is much straighter.

b. $H_o : \rho_1 = 0$ will be rejected in favor of $H_a : \rho \neq 0$ at level .01 if
$|t| \geq t_{.005,8} = 3.355$.

$\Sigma x_i = 864$, $\Sigma x_i^2 = 78{,}142$, $\Sigma y_i = 138.0$, $\Sigma y_i^2 = 1959.1$ and

$\Sigma x_i y_i = 12{,}322.4$, so $r = \dfrac{3992}{(186.8796)(23.3880)} = .913$ and

$t = \dfrac{.913(2.8284)}{.4080} = 6.33 \geq 3.355$, so reject H_o. There does appear to be a linear relationship.

67.

    a. Because p-value = .00032 < α = .001, H_o should be rejected at this significance level.

    b. Not necessarily. For this n, the test statistic t has approximately a standard normal distribution when $H_o : \rho_1 = 0$ is true, and a p-value of .00032 corresponds to $z = 3.60$ (or –3.60). Solving $3.60 = \dfrac{r\sqrt{498}}{\sqrt{1}} - r^2$ for r yields r = .159. This r suggests only a weak linear relationship between x and y, one that would typically have little practical import.

    c. $t = 2.20 \geq t_{.025,9998} = 1.96$, so H_o is rejected in favor of H_a. The value t = 2.20 is statistically significant -- it cannot be attributed just to sampling variability in the case $\rho = 0$. But with this n, r = .022 implies $\rho = .022$, which in turn shows an extremely weak linear relationship.

Supplementary Exercises

69.

    a. The test statistic value is $t = \dfrac{\hat{\beta}_1 - 1}{s_{\hat{\beta}_1}}$, and H_o will be rejected if either $t \geq t_{.025,11} = 2.201$ or $t \leq -2.201$. With $\Sigma x_i = 243, \Sigma x_i^2 = 5965, \Sigma y_i = 241, \Sigma y_i^2 = 5731$ and $\Sigma x_i y_i = 5805$, $\hat{\beta}_1 = .913819$, $\hat{\beta}_0 = 1.457072$, $SSE = 75.126$, $s = 2.613$, and $s_{\hat{\beta}_1} = .0693$, $t = \dfrac{.9138 - 1}{.0693} = -1.24$. Because –1.24 is neither ≤ -2.201 nor ≥ 2.201, H_o cannot be rejected. It is plausible that $\beta_1 = 1$.

    b. $r = \dfrac{16{,}902}{(136)(128.15)} = .970$

71.

a. $r^2 = .5073$

b. $r = +\sqrt{r^2} = \sqrt{.5073} = .7122$ (positive because $\hat{\beta}_1$ is positive.)

c. We test test $H_0 : \beta_1 = 0$ vs $H_0 : \beta_1 \neq 0$. The test statistic t = 3.93 gives p-value = .0013, which is < .01, the given level of significance, therefore we reject H_0 and conclude that the model is useful.

d. We use a 95% CI for $\mu_{Y \cdot 50}$. $\hat{y}_{(50)} = .787218 + .007570(50) = 1.165718$,

$t_{.025,15} = 2.131$, s = "Root MSE" = .020308, so

$$s_{\hat{y}_{(50)}} = .20308\sqrt{\frac{1}{17} + \frac{17(50 - 42.33)^2}{17(41,575) - (719.60)^2}} = .051422 . \text{ The interval}$$

is , then,

$$1.165718 \pm 2.131(.051422) = 1.165718 \pm .109581 = (1.056137, 1.275299)$$

e. $\hat{y}_{(30)} = .787218 + .007570(30) = 1.0143$. The residual is

$y - \hat{y} = .80 - 1.0143 = -.2143$.

73.

a. $n = 9, \Sigma x_i = 228, \Sigma x_i^2 = 5958, \Sigma y_i = 93.76, \Sigma y_i^2 = 982.2932$ and

$\Sigma x_i y_i = 2348.15$, giving $\hat{\beta}_1 = \dfrac{-243.93}{1638} = -.148919$,

$\hat{\beta}_0 = 14.190392$, and the equation $\hat{y} = 14.19 - (.1489)x$.

b. β_1 is the expected increase in load associated with a one-day age increase (so a negative value of β_1 corresponds to a decrease). We wish to test

$H_0 : \beta_1 = -.10$ vs. $H_0 : \beta_1 < -.10$ (the alternative contradicts prior belief).

H_0 will be rejected at level .05 if $t = \dfrac{\hat{\beta}_1 - (-.10)}{s_{\hat{\beta}_1}} \le -t_{.05,7} = -1.895$. With

SSE = 1.4862, s = .4608, and $s_{\hat{\beta}_1} = \dfrac{.4608}{\sqrt{182}} = .0342$. Thus

$t = \dfrac{-.1489 + 1}{.0342} = -1.43$. Because -1.43 is not ≤ -1.895, do not reject H_0.

c. $\Sigma x_i = 306, \Sigma x_i^2 = 7946$, so $\sum (x_i - \bar{x})^2 = 7946 - \dfrac{(306)^2}{12} = 143$ here,

as contrasted with 182 for the given 9 x_i's. Even though the sample size for the proposed x values is larger, the original set of values is preferable.

d. $(t_{.025,7})(s)\sqrt{\dfrac{1}{9} + \dfrac{9(28 - 25.33)^2}{1638}} = (2.365)(.4608)(.3877) = .42$, and

$\hat{\beta}_0 + \hat{\beta}_1(28) = 10.02$, so the 95% CI is $10.02 \pm .42 = (9.60, 10.44)$.

75.

a. The plot suggests a strong linear relationship between x and y.

b. $n = 9$, $\Sigma x_i = 1797$, $\Sigma x_i^2 = 4334.41$, $\Sigma y_i = 7.28$, $\Sigma y_i^2 = 7.4028$ and

$\Sigma x_i y_i = 178.683$, so $\hat{\beta}_1 = \dfrac{299.931}{6717.6} = .04464854$, $\hat{\beta}_0 = -.08259353$,

and the equation of the estimated line is $\hat{y} = -.08259 - (.044649)x$.

c. $SSE = 7.4028 - (-601281) - 7.977935 = .026146$,

$SST = 7.4028 - \dfrac{(7.28)^2}{9} = .026146, = 1.5141$, and

$r^2 = 1 - \dfrac{SSE}{SST} = .983$, so 93.8% of the observed variation is "explained."

d. $\hat{y}_4 = -.08259 - (.044649)(19.1) = .7702$, and

$y_4 - \hat{y}_4 = .68 - .7702 = -.0902$.

e. $s = .06112$, and $s_{\hat{\beta}_1} = \dfrac{.06112}{\sqrt{746.4}} = .002237$, so the value of t for testing

$H_0 : \beta_1 = 0$ vs $H_0 : \beta_1 \neq 0$ is $t = \dfrac{.044649}{.002237} = 19.96$. From Table A.5,

$t_{.0005,7} = 5.408$, so $p - value < 2(.0005) = .001$. There is strong evidence

for a useful relationship.

f. A 95% CI for β_1 is $.044649 \pm (2.365)(.002237) = .044649 \pm .005291$

$= (.0394, .0499)$.

g. A 95% CI for $\beta_0 + \beta_1(20)$ is $.810 \pm (2.365)(.002237)(.3333356)$

$= .810 \pm .048 = (.762, .858)$

77. $SSE = \Sigma y^2 - \hat{\beta}_0 \Sigma y - \hat{\beta}_1 \Sigma xy$. Substituting $\hat{\beta}_0 = \dfrac{\Sigma y - \hat{\beta}_1 \Sigma x}{n}$, SSE becomes

$$SSE = \Sigma y^2 - \frac{\Sigma y \left(\Sigma y - \hat{\beta}_1 \Sigma x \right)}{n} - \hat{\beta}_1 \Sigma xy = \Sigma y^2 - \frac{(\Sigma y)^2}{n} + \frac{\hat{\beta}_1 \Sigma x \Sigma y}{n} - \hat{\beta}_1 \Sigma xy$$

$$= \left[\Sigma y^2 - \frac{(\Sigma y)^2}{n} \right] - \hat{\beta}_1 \left[\Sigma xy - \frac{\Sigma x \Sigma y}{n} \right] = S_{yy} - \hat{\beta}_1 S_{xy}, \text{ as desired.}$$

79.

a. With $S_{xx} = \sum (x_i - \bar{x})^2$, $S_{yy} = \sum (y_i - \bar{y})^2$, note that $\dfrac{s_y}{s_x} = \sqrt{\dfrac{S_{yy}}{S_{xx}}}$ (since

the factor n-1 appears in both the numerator and denominator, so cancels). Thus

$$y = \hat{\beta}_0 + \hat{\beta}_1 x = \bar{y} + \hat{\beta}_1 (x - \bar{x}) = \bar{y} + \frac{S_{xy}}{S_{xx}} (x - \bar{x}) = \bar{y} + \sqrt{\frac{S_{yy}}{S_{xx}}} \cdot \frac{S_{xy}}{\sqrt{S_{xx} S_{yy}}} (x - \bar{x})$$

$$= \bar{y} + \frac{s_y}{s_x} \cdot r \cdot (x - \bar{x}), \text{ as desired.}$$

b. By .573 s.d.'s above, (above, since r < 0) or (since $s_y = 4.3143$) an amount 2.4721 above.

81. Using the notation of the exercise above, $SST = s_{yy}$ and $SSE = s_{yy} - \hat{\beta}_1 s_{xy}$

$$= s_{yy} - \frac{s_{xy}^2}{s_{xx}}, \text{ so } 1 - \frac{SSE}{SST} = 1 - \frac{s_{yy} - \dfrac{s_{xy}^2}{s_{xx}}}{s_{yy}} = \frac{s_{xy}^2}{s_{xx} s_{yy}} = r^2, \text{ as desired.}$$

83.   Using Minitab, we create a scatterplot to see if a linear regression model is appropriate.

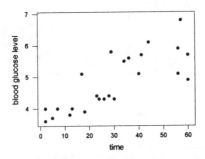

A linear model is reasonable; although it appears that the variance in y gets larger as x increases. The Minitab output follows:

```
The regression equation is
blood glucose level = 3.70 + 0.0379 time

Predictor        Coef        StDev           T          P
Constant       3.6965       0.2159       17.12      0.000
time         0.037895     0.006137        6.17      0.000

S = 0.5525      R-Sq = 63.4%      R-Sq(adj) = 61.7%

Analysis of Variance

Source           DF         SS          MS          F          P
Regression        1      11.638      11.638      38.12      0.000
Residual Error   22       6.716       0.305
Total            23      18.353
```

The coefficient of determination of 63.4% indicates that only a moderate percentage of the variation in y can be explained by the change in x. A test of model utility indicates that time is a significant predictor of blood glucose level. (t = 6.17, p = 0.0). A point estimate for blood glucose level when time = 30 minutes is 4.833%. We would expect the average blood glucose level at 30 minutes to be between 4.599 and 5.067, with 95% confidence.

85. For the second boiler, $n = 19$, $\Sigma x_i = 125$, $\Sigma y_i = 472.0$, $\Sigma x_i^2 = 3625$, $\Sigma y_i^2 = 37,140.82$, and $\Sigma x_i y_i = 9749.5$, giving $\hat{\gamma}_1 = $ estimated slope

$$= \frac{-503}{6125} = -.0821224, \ \hat{\gamma}_0 = 80.377551, \ SSE_2 = 3.26827,$$

$SSx_2 = 1020.833$. For boiler #1, $n = 8$, $\hat{\beta}_1 = -.1333$, $SSE_1 = 8.733$, and

$SSx_1 = 1442.875$. Thus $\hat{\sigma}^2 = \dfrac{8.733 + 3.286}{10} = 1.2$, $\hat{\sigma} = 1.095$, and

$$t = \frac{-.1333 + .0821}{1.095\sqrt{\frac{1}{1442.875} + \frac{1}{1020.833}}} = \frac{-.0512}{.0448} = -1.14.$$ $t_{.025,10} = 2.228$ and -1.14 is

neither ≥ 2.228 nor ≤ -2.228, so H_o is not rejected. It is plausible that $\beta_1 = \gamma_1$.

Chapter 12: Simple Linear Regression and Correlation

CHAPTER 13

Section 13.1

1.

   a. $\bar{x} = 15$ and $\sum (x_j - \bar{x})^2 = 250$, so s.d. of $Y_i - \hat{Y}_i$ is

$$10\sqrt{1 - \frac{1}{5} - \frac{(x_i - 15)^2}{250}} = 6.32,\ 8.37,\ 8.94,\ 8.37,\ \text{and}\ 6.32\ \text{for}\ i = 1, 2, 3, 4, 5.$$

   b. Now $\bar{x} = 20$ and $\sum (x_i - \bar{x})^2 = 1250$, giving standard deviations 7.87, 8.49, 8.83, 8.94, and 2.83 for $i = 1, 2, 3, 4, 5$.

   c. The deviation from the estimated line is likely to be much smaller for the observation made in the experiment of **b** for x = 50 than for the experiment of **a** when x = 25. That is, the observation (50, Y) is more likely to fall close to the least squares line than is (25, Y).

3.

   a. This plot indicates there are no outliers, the variance of ε is reasonably constant, and the ε are normally distributed. A straight-line regression function is a reasonable choice for a model.

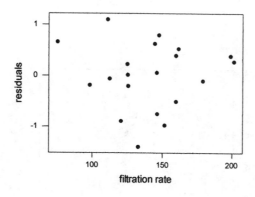

253

b. We need $S_{xx} = \sum (x_i - \bar{x})^2 = 415,914.85 - \dfrac{(2817.9)^2}{20} = 18,886.8295$. Then

each e_i^* can be calculated as follows: $e_i^* = \dfrac{e_i}{.4427\sqrt{1 + \dfrac{1}{20} + \dfrac{(x_i - 140.895)^2}{18,886.8295}}}$.

The table below shows the values:

standardized residuals	e/e_i^*	standardized residuals	e/e_i^*
-0.31064	0.644053	0.6175	0.64218
-0.30593	0.614697	0.09062	0.64802
0.4791	0.578669	1.16776	0.565003
1.2307	0.647714	-1.50205	0.646461
-1.15021	0.648002	0.96313	0.648257
0.34881	0.643706	0.019	0.643881
-0.09872	0.633428	0.65644	0.584858
-1.39034	0.640683	-2.1562	0.647182
0.82185	0.640975	-0.79038	0.642113
-0.15998	0.621857	1.73943	0.631795

Notice that if $e_i^* \approx e/s$, then $e/e_i^* \approx s$. All of the e/e_i^*'s range between .57 and .65, which are close to s.

c. This plot looks very much the same as the one in part a.

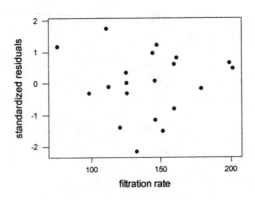

5.

a. 97.7% of the variation in ice thickness can be explained by the linear relationship between it and elapsed time. Based on this value, it appears that a linear model is reasonable.

b. The residual plot shows a curve in the data, so perhaps a non-linear relationship exists. One observation (5.5, -3.14) is extreme.

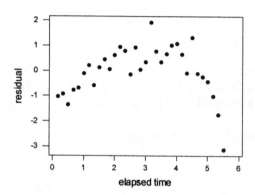

7.

a.

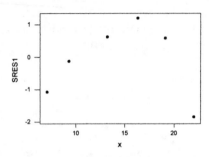

There is an obvious curved pattern in the scatter plot, which suggests that a simple linear model will not provide a good fit.

b. The $\hat{y}'s$, e's, and e*'s are given below:

x	y	$\hat{y}$	e	e*
0	110	126.6	-16.6	-1.55
2	123	113.3	9.7	.68
4	119	100.0	19.0	1.25
6	86	86.7	-.7	-.05
8	62	73.4	-11.4	-1.06

9.   Both a scatter plot and residual plot (based on the simple linear regression model) for the first data set suggest that a simple linear regression model is reasonable, with no pattern or influential data points which would indicate that the model should be modified. However, scatter plots for the other three data sets reveal difficulties.

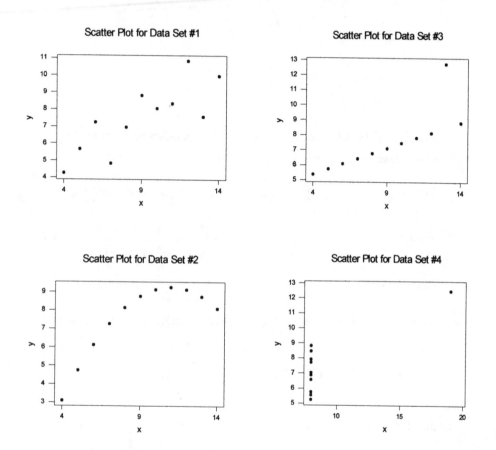

For data set #2, a quadratic function would clearly provide a much better fit. For data set #3, the relationship is perfectly linear except one outlier, which has obviously greatly influenced the fit even though its x value is not unusually large or small. The signs of the residuals here (corresponding to increasing x) are + + + + - - - - - + -, and a residual plot would reflect this pattern and suggest a careful look at the chosen model. For data set #4 it is clear that the slope of the least squares line has been determined entirely by the outlier, so this point is extremely influential (and its x value does lie far from the remaining ones).

11.

a. $Y_i - \hat{Y}_i = Y_i - \bar{Y} - \hat{\beta}_1(x_i - \bar{x}) = Y_i - \dfrac{1}{n}\sum_j Y_j - \dfrac{(x_i - \bar{x})\sum_j(x_j - \bar{x})Y_j}{\sum_j(x_j - \bar{x})^2} = \sum_j c_j Y_j$,

where $c_j = 1 - \dfrac{1}{n} - \dfrac{(x_i - \bar{x})^2}{n\Sigma(x_j - \bar{x})^2}$ for $j = i$ and

$c_j = 1 - \dfrac{1}{n} - \dfrac{(x_i - \bar{x})(x_j - \bar{x})}{\Sigma(x_j - \bar{x})^2}$ for $j \neq i$. Thus

$Var(Y_i - \hat{Y}_i) = \Sigma Var(c_j Y_j)$ (since the Y_j's are independent) $= \sigma^2 \Sigma c_j^2$

which, after some algebra, gives equation (13.2).

b. $\sigma^2 = Var(Y_i) = Var(\hat{Y}_i + (Y_i - \hat{Y}_i)) = Var(\hat{Y}_i) + Var(Y_i - \hat{Y}_i)$, so

$Var(Y_i - \hat{Y}_i) = \sigma^2 - Var(\hat{Y}_i) = \sigma^2 - \sigma^2\left[\dfrac{1}{n} + \dfrac{(x_i - \bar{x})^2}{n\Sigma(x_j - \bar{x})^2}\right]$, which is

exactly (13.2).

c. As x_i moves further from $\bar{x}$, $(x_i - \bar{x})^2$ grows larger, so $Var(\hat{Y}_i)$ increases (since $(x_i - \bar{x})^2$ has a positive sign in $Var(\hat{Y}_i)$), but $Var(Y_i - \hat{Y}_i)$ decreases (since $(x_i - \bar{x})^2$ has a negative sign).

13. The distribution of any particular standardized residual is also a t distribution with n – 2 d.f., since e_i^* is obtained by taking standard normal variable $\dfrac{(Y_i - \hat{Y}_i)}{(\sigma_{Y_i - \hat{Y}})}$ and substituting the estimate of σ in the denominator (exactly as in the predicted value case). With E_i^* denoting the ith standardized residual as a random variable, when n = 25 E_i^* has a t distribution with 23 d.f. and $t_{.01,23} = 2.50$, so P(E_i^* outside (-2.50, 2.50)) = $P(E_i^* \geq 2.50) + P(E_i^* \leq -2.50) = .01 + .01 = .02$.

Section 13.2

15.

    a.

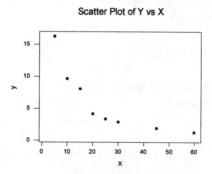

Scatter Plot of Y vs X

The points have a definite curved pattern. A linear model would not be appropriate.

    b. In this plot we have a strong linear pattern.

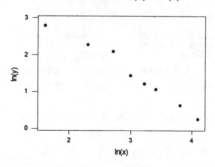

Scatter Plot of ln(Y) vs ln(X)

    c. The linear pattern in **b** above would indicate that a transformed regression using the natural log of both x and y would be appropriate. The probabilistic model is then $y = \alpha x^{\beta} \cdot \varepsilon$. (The power function with an error term!)

d. A regression of ln(y) on ln(x) yields the equation
$\ln(y) = 4.6384 - 1.04920 \ln(x)$. Using Minitab we can get a P.I. for y
when x = 20 by first transforming the x value: ln(20) = 2.996. The computer
generated 95% P.I. for ln(y) when ln(x) = 2.996 is (1.1188,1.8712). We must
now take the antilog to return to the original units of Y:
$\left(e^{1.1188}, e^{1.8712}\right) = (3.06, 6.50)$.

e. A computer generated residual analysis:

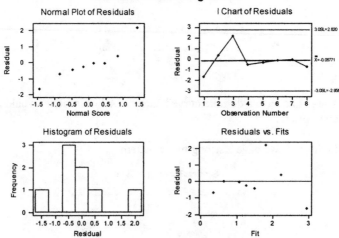

Residual Model Diagnostics

Looking at the residual vs. fits (bottom right), one standardized residual,
corresponding to the third observation, is a bit large. There are only two positive
standardized residuals, but two others are essentially 0. The patterns in the
residual plot and the normal probability plot (upper left) are marginally
acceptable.

17.

a.

$\Sigma x_i' = 15.501$, $\Sigma y_i' = 13.352$, $\Sigma x_i'^2 = 20.228$, $\Sigma y_i'^2 = 16.572$,

$\Sigma x_i' \ y_i' = 18.109$, from which $\hat{\beta}_1 = 1.254$ and $\hat{\beta}_0 = -.468$ so

$\hat{\beta} = \hat{\beta}_1 = 1.254$ and $\hat{\alpha} = e^{-.468} = .626$.

b. The plots give strong support to this choice of model; in addition, $r^2 = .960$ for the transformed data.

c. SSE $= .11536$ (computer printout), $s = .1024$, and the estimated sd of $\hat{\beta}_1$ is

$.0775$, so $t = \dfrac{1.25 - 1.33}{.0775} = -1.07$. Since -1.07 is not $\le -t_{.05,11} = -1.796$,

H_o cannot be rejected in favor of H_a.

d. The claim that $\mu_{Y.5} = 2\mu_{Y.2.5}$ is equivalent to $\alpha \cdot 5^{\beta} = 2\alpha(2.5)^{\beta}$, or that

$\beta = 1$. Thus we wish test $H_o : \beta_1 = 1$ vs. $H_a : \beta_1 \ne 1$. With

$t = \dfrac{1 - 1.33}{.0775} = -4.30$ and RR $-t_{.005,11} \le -3.106$, H_o is rejected at level $.01$

since $-4.30 \le -3.106$.

19.

a. No, there is definite curvature in the plot.

b. $Y' = \beta_0 + \beta_1(x') + \varepsilon$ where $x' = \dfrac{1}{temp}$ and $y' = \ln(lifetime)$. Plotting y' vs. x' gives a plot which has a pronounced linear appearance (and in fact r^2 = .954 for the straight line fit).

c. $\Sigma x'_i = .082273$, $\Sigma y'_i = 123.64$, $\Sigma x'^2_i = .00037813$, $\Sigma y'^2_i = 879.88$, $\Sigma x'_i \, y'_i = .57295$, from which $\hat{\beta}_1 = 3735.4485$ and $\hat{\beta}_0 = -10.2045$ (values read from computer output). With x = 220, $x' = .00445$ so $\hat{y}' = -10.2045 + 3735.4485(.00445) = 6.7748$ and thus $\hat{y} = e^{\hat{y}'} = 875.50$.

d. For the transformed data, SSE = 1.39857, and $n_1 = n_2 = n_3 = 6$, $\bar{y}'_{1.} = 8.44695$, $\bar{y}'_{2.} = 6.83157$, $\bar{y}'_{3.} = 5.32891$, from which SSPE = 1.36594, SSLF = .02993, $f = \dfrac{.02993/1}{1.36594/15} = .33$. Comparing this to $F_{.01,1,15} = 8.68$, it is clear that H_o cannot be rejected.

21.

a. The suggested model is $Y = \beta_0 + \beta_1(x') + \varepsilon$ where $x' = \dfrac{10^4}{x}$. The summary quantities are $\Sigma x'_i = 159.01$, $\Sigma y_i = 121.50$, $\Sigma x'^2_i = 4058.8$, $\Sigma y^2_i = 1865.2$, $\Sigma x'_i \, y_i = 2281.6$, from which $\hat{\beta}_1 = -.1485$ and $\hat{\beta}_0 = 18.1391$, and the estimated regression function is $y = 18.1391 - \dfrac{1485}{x}$.

b. $x = 500 \Rightarrow \hat{y} = 18.1391 - \dfrac{1485}{500} = 15.17$.

23. $Var(Y) = Var(\alpha e^{\beta x} \cdot \varepsilon) = \left[\alpha e^{\beta x}\right]^2 \cdot Var(\varepsilon) = \alpha^2 e^{2\beta x} \cdot \tau^2$ where we have set $Var(\varepsilon) = \tau^2$. If $\beta > 0$, this is an increasing function of x so we expect more spread in y for large x than for small x, while the situation is reversed if $\beta < 0$. It is important to realize that a scatter plot of data generated from this model will not spread out uniformly about the exponential regression function throughout the range of x values; the spread will only be uniform on the transformed scale. Similar results hold for the multiplicative power model.

25.   The point estimate of β_1 is $\hat{\beta}_1 = .17772$, so the estimate of the odds ratio is

$e^{\hat{\beta}_1} = e^{.17772} \approx 1.194$. That is , when the amount of experience increases by one year (i.e. a one unit increase in x), we estimate that the odds ratio increase by about 1.194. The z value of 2.70 and its corresponding p-value of .007 imply that the null hypothesis $H_0 : \beta_1 = 0$ can be rejected at any of the usual significance levels (e.g., .10, .05, .025, .01). Therefore, there is clear evidence that β_1 is not zero, which means that experience does appear to affect the likelihood of successfully performing the task. This is consistent with the confidence interval (1.05, 1.36) for the odds ratio given in the printout, since this interval does not contain the value 1. A graph of $\hat{\pi}$ appears below.

27.

a. A scatter plot of the data indicated a quadratic regression model might be appropriate.

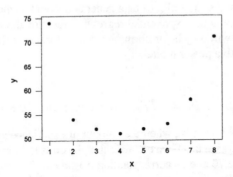

b. $\hat{y} = 84.482 - 15.875(6) + 1.7679(6)^2 = 52.88$; residual =
$y_6 - \hat{y}_6 = 53 - 52.88 = .12$;

c. $SST = \Sigma y_i^2 - \dfrac{(\Sigma y_i)^2}{n} = 586.88$, so $R^2 = 1 - \dfrac{61.77}{586.88} = .895$.

d. The first two residuals are the largest, but they are both within the interval (-2, 2). Otherwise, the standardized residual plot does not exhibit any troublesome features. For the Normal Probability Plot:

Residual	Zth percentile
-1.95	-1.53
-.66	-.89
-.25	-.49
.04	-.16
.20	.16
.58	.49
.90	.89
1.91	1.53

The normal probability plot does not exhibit any troublesome features.

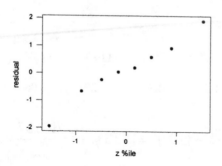

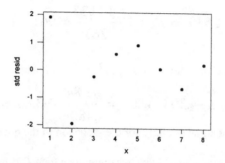

e. $\hat{\mu}_{Y \cdot 6} = 52.88$ (from **b**) and $t_{.025, n-3} = t_{.025, 5} = 2.571$, so the C.I. is
$52.88 \pm (2.571)(1.69) = 52.88 \pm 4.34 = (48.54, 57.22)$.

f. SSE = 61.77 so $s^2 = \dfrac{61.77}{5} = 12.35$ and $\sqrt{12.35 + (1.69)^2} = 3.90$. The
P.I. is $52.88 \pm (2.571)(3.90) = 52.88 \pm 10.03 = (42.85, 62.91)$.

Section 13.3

29.

a. From computer output:

$$\hat{y}: \quad 111.89 \quad 120.66 \quad 114.71 \quad 94.06 \quad 58.69$$

$$y - \hat{y}: \quad -1.89 \quad 2.34 \quad 4.29 \quad -8.06 \quad 3.31$$

Thus $SSE = (-1.89)^2 + \ldots + (3.31)^2 = 103.37$, $s^2 = \dfrac{103.37}{2} = 51.69$,

$s = 7.19$.

b. $SST = \Sigma y_i^2 - \dfrac{(\Sigma y_i)^2}{n} = 2630$, so $R^2 = 1 - \dfrac{103.37}{2630} = .961$.

c. $H_0: \beta_2 = 0$ will be rejected in favor of $H_a: \beta_2 \neq 0$ if either

$t \geq t_{.025,2} = 4.303$ or if $t \leq -4.303$. With $t = \dfrac{-1.84}{.480} = -3.83$, H_0 cannot be

rejected; the data does not argue strongly for the inclusion of the quadratic term.

d. To obtain joint confidence of at least 95%, we compute a 98% C.I. for each coefficient using $t_{.01,2} = 6.965$. For β_1 the C.I. is $8.06 \pm (6.965)(4.01)$

$= (-19.87, 35.99)$ (an extremely wide interval), and for β_2 the C.I. is

$-1.84 \pm (6.965)(.480) = (-5.18, 1.50)$.

e. $t_{.05,2} = 2.920$ and $\hat{\beta}_0 + 4\hat{\beta}_1 + 16\hat{\beta}_2 = 114.71$, so the C.I. is

$114.71 \pm (2.920)(5.01) = 114.71 \pm 14.63 = (100.08, 129.34)$.

f. If we knew $\hat{\beta}_0, \hat{\beta}_1, \hat{\beta}_2$, the value of x which maximizes $\hat{\beta}_0 + \hat{\beta}_1 x + \hat{\beta}_2 x^2$ would be obtained by setting the derivative of this to 0 and solving:

$\beta_1 + 2\beta_2 x = 0 \Rightarrow x = -\dfrac{\beta_1}{2\beta_2}$. The estimate of this is $x = -\dfrac{\hat{\beta}_1}{2\hat{\beta}_2} = 2.19$.

31.

a. Using Minitab, the regression equation is $y = 13.6 + 11.4x - 1.72x^2$.

b. Again, using Minitab, the predicted and residual values are:

$\hat{y}$:	23.327	23.327	29.587	31.814	31.814	31.814	20.317
$y - \hat{y}$:	-.327	1.173	1.587	.914	.186	1.786	-.317

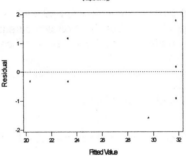

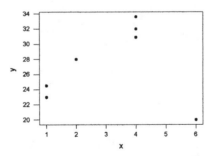

The residual plot is consistent with a quadratic model (no pattern which would suggest modification), but it is clear from the scatter plot that the point (6, 20) has had a great influence on the fit – it is the point which forced the fitted quadratic to have a maximum between 3 and 4 rather than, for example, continuing to curve slowly upward to a maximum someplace to the right of x = 6.

c. From Minitab output, $s^2 = \text{MSE} = 2.040$, and $R^2 = 94.7\%$. The quadratic model thus explains 94.7% of the variation in the observed y's, which suggests that the model fits the data quite well.

d. $\sigma^2 = Var(\hat{Y}_i) + Var\left(Y_i - \hat{Y}_i\right)$ suggests that we can estimate $Var\left(Y_i - \hat{Y}_i\right)$ by $s^2 - s_{\hat{y}}^2$ and then take the square root to obtain the estimated standard deviation of each residual. This gives $\sqrt{2.040 - (.955)^2} = 1.059$, (and similarly for all points) 10.59, 1.236, 1.196, 1.196, 1.196, and .233 as the estimated std dev's of the residuals. The standardized residuals are then computed as $\dfrac{-.327}{1.059} = -.31$, (and similarly) 1.10, -1.28, -.76, .16, 1.49, and –1.28, none of which are unusually large. (Note: Minitab regression output can produce these values.) The resulting residual plot is virtually identical to the plot of **b**.

$\dfrac{y - \hat{y}}{s} = \dfrac{-.327}{1.426} = -.229 \neq -.31$, so standardizing using just s would not yield the correct standardized residuals.

e. $Var(Y_f) + Var(\hat{Y}_f)$ is estimated by $2.040 + (.777)^2 = 2.638$, so $s_{y_f + \hat{y}_f} = \sqrt{2.638} = 1.624$. With $\hat{y} = 31.81$ and $t_{.05,4} = 2.132$, the desired P.I. is $31.81 \pm (2.132)(1.624) = (28.35, 35.27)$.

33.

a. $\bar{x} = 20$ and $s_x = 10.8012$ so $x' = \dfrac{x-20}{10.8012}$. For x = 20, $x' = 0$, and

$\hat{y} = \hat{\beta}_0^* = .9671$. For x = 25, $x' = .4629$, so

$\hat{y} = .9671 - .0502(.4629) - .0176(.4629)^2 + .0062(.4629)^3 = .9407$.

b. $\hat{y} = .9671 - .0502\left(\dfrac{x-20}{10.8012}\right) - .0176\left(\dfrac{x-20}{10.8012}\right)^2 + .0062\left(\dfrac{x-20}{10.8012}\right)^3$

$.00000492x^3 - .000446058x^2 + .007290688x + .96034944$.

c. $t = \dfrac{.0062}{.0031} = 2.00$. We reject H_o if either $t \geq t_{.025,n-4} = t_{.025,3} = 3.182$ or if

$t \leq -3.182$. Since 2.00 is neither ≥ 3.182 nor ≤ -3.182, we cannot reject
H_o; the cubic term should be deleted.

d. $SSE = \Sigma(y_i - \hat{y}_i)$ and the $\hat{y}_i$'s are the same from the standardized as from
the unstandardized model, so SSE, SST, and R^2 will be identical for the two
models.

e. $\Sigma y_i^2 = 6.355538$, $\Sigma y_i = 6.664$, so SST = .011410. For the quadratic model
$R^2 = .987$ and for the cubic model, $R^2 = .994$; The two R^2 values are very close,
suggesting intuitively that the cubic term is relatively unimportant.

35. $Y' = \ln(Y) = \ln\alpha + \beta x + \gamma x^2 + \ln(\varepsilon) = \beta_0 + \beta_1 x + \beta_2 x^2 + \varepsilon'$ where
$\varepsilon' = \ln(\varepsilon)$, $\beta_0 = \ln(\alpha)$, $\beta_1 = \beta$, and $\beta_2 = \gamma$. That is, we should fit a
quadratic to $(x, \ln(y))$. The resulting estimated quadratic (from computer output) is
$2.00397 + .1799x - .0022x^2$, so $\hat{\beta} = .1799$, $\hat{\gamma} = -.0022$, and
$\hat{\alpha} = e^{2.0397} = 7.6883$. (The ln(y)'s are 3.6136, 4.2499, 4.6977, 5.1773, and
5.4189, and the summary quantities can then be computed as before.)

Section 13.4

37.

    a.   The mean value of y when $x_1 = 50$ and $x_2 = 3$ is

$$\mu_{y \cdot 50,3} = -.800 + .060(50) + .900(3) = 4.9 \text{ hours.}$$

    b.   When the number of deliveries (x_2) is held fixed, then average change in travel time associated with a one-mile (i.e. one unit) increase in distance traveled (x_1) is .060 hours. Similarly, when distance traveled (x_1) is held fixed, then the average change in travel time associated with on extra delivery (i.e., a one unit increase in x_2) is .900 hours.

    c.   Under the assumption that y follows a normal distribution, the mean and standard deviation of this distribution are 4.9 (because $x_1 = 50$ and $x_2 = 3$) and $\sigma = .5$ (since the standard deviation is assumed to be constant regardless of the values of x_1 and x_2). Therefore $P(y \le 6) = P\left(z \le \dfrac{6 - 4.9}{.5} \right) = P(z \le 2.20) = .9861$.

    That is, in the long run, about 98.6% of all days will result in a travel time of at most 6 hours.

39.

    a.   For $x_1 = 2$, $x_2 = 8$ (remember the units of x_2 are in 1000,s) and $x_3 = 1$ (since the outlet has a drive-up window) the average sales are

$$\hat{y} = 10.00 - 1.2(2) + 6.8(8) + 15.3(1) = 77.3 \text{ (i.e., \$77,300).}$$

    b.   For $x_1 = 3$, $x_2 = 5$, and $x_3 = 0$ the average sales are

$$\hat{y} = 10.00 - 1.2(3) + 6.8(5) + 15.3(0) = 40.4 \text{ (i.e., \$40,400).}$$

    c.   When the number of competing outlets (x_1) and the number of people within a 1-mile radius (x_2) remain fixed, the sales will increase by \$15,300 when an outlet has a drive-up window.

41. $H_0 : \beta_1 = \beta_2 = ... = \beta_6 = 0$ vs. H_a: at least one among $\beta_1, ..., \beta_6$ is not zero. The

test statistic is $F = \dfrac{R^2/k}{(1-R^2)/(n-k-1)}$. H_0 will be rejected if $f \geq F_{.05,6,30} = 2.42$.

$f = \dfrac{.83/6}{(1-.83)/30} = 24.41$. Because $24.41 \geq 2.42$, H_0 is rejected and the model is judged

useful.

43.

   a. The coefficient of multiple determination is $R^2 = 78\%$. So 78% of the
           observed variation in surface area can be attributed to the stated approximate
           relationship between surface area and the predictors.

   b. $x_1 = 2.6$, $x_2 = 250$, and $x_1x_2 = (2.6)(250) = 650$, so
           $\hat{y} = 185.49 - 45.97(2.6) - 0.3015(250) + 0.0888(650) = 48.313$

   c. No, it is not legitimate to interpret β_1 in this way. It is not possible to increase
           by 1 unit the cobalt content, x_1, while keeping the interaction predictor, x_3, fixed.
           When x_1 changes, so does x_3, since $x_3 = x_1x_2$.

   d. Yes, there appears to be a useful linear relationship between y and the predictors.
           We determine this by observing that the p-value corresponding to the model
           utility test is $< .0001$ (F test statistic $= 18.924$).

   e. We wish to test $H_0 : \beta_3 = 0$ vs. $H_a : \beta_3 \neq 0$. The test statistic is t=3.496,
           with a corresponding p-value of .0030. Since the p-value is $<$ alpha $= .01$, we
           reject H_0 and conclude that the interaction predictor does provide useful
           information about y.

   f. A 95% C.I. for the mean value of surface area under the stated circumstances
           requires the following quantities:
           $\hat{y} = 185.49 - 45.97(2) - 0.3015(500) + 0.0888(2)(500) = 31.598$.
           Next, $t_{.025,16} = 2.120$, so the 95% confidence interval is
           $$31.598 \pm (2.120)(4.69) = 31.598 \pm 9.9428 = (21.6552, 41.5408)$$

45.

a. The appropriate hypotheses are $H_0 : \beta_1 = \beta_2 = \beta_3 = \beta_4 = 0$ vs. H_a : at least one $\beta_i \neq 0$. The test statistic is

$$f = \frac{R^2/k}{(1-R^2)/(n-k-1)} = \frac{.946/4}{(1-.946)/20} = 87.6 \geq 7.10 = F_{.001,4,20} \text{ (the smallest available}$$

significance level from Table A.9), so we can reject H_0 at any significance level. We conclude that at least one of the four predictor variables appears to provide useful information about tenacity.

b. The adjusted R^2 value is $1 - \dfrac{n-1}{n-(k+1)}\left(\dfrac{SSE}{SST}\right) = 1 - \dfrac{n-1}{n-(k+1)}\left(1-R^2\right)$

$= 1 - \dfrac{24}{20}(1-.946) = .935$, which does not differ much from $R^2 = .946$.

c. The estimated average tenacity when $x_1 = 16.5$, $x_2 = 50$, $x_3 = 3$, and $x_4 = 5$ is
$\hat{y} = 6.121 - .082x + .113x + .256x - .219x$
$\hat{y} = 6.121 - .082(16.5) + .113(50) + .256(3) - .219(5) = 10.091$. For a 99% C.I.,
$t_{.005,20} = 2.845$, so the interval is $10.091 \pm 2.845(.350) = (9.095, 11.087)$.

Therefore, when the four predictors are as specified in this problem, the true average tenacity is estimated to be between 9.095 and 11.087.

47.

    a. For a 1% increase in the percentage plastics, we would expect a 28.9 kcal/kg increase in energy content. Also, for a 1% increase in the moisture, we would expect a 37.4 kcal/kg decrease in energy content.

    b. The appropriate hypotheses are $H_0 : \beta_1 = \beta_2 = \beta_3 = \beta_4 = 0$ vs. H_a : at least one $\beta_i \neq 0$. The value of the F test statistic is 167.71, with a corresponding p-value that is extremely small. So, we reject H_o and conclude that at least one of the four predictors is useful in predicting energy content, using a linear model.

    c. $H_0 : \beta_3 = 0$ vs. $H_a : \beta_3 \neq 0$. The value of the t test statistic is t = 2.24, with a corresponding p-value of .034, which is less than the significance level of .05. So we can reject H_o and conclude that percentage garbage provides useful information about energy consumption, given that the other three predictors remain in the model.

    d. $\hat{y} = 2244.9 + 28.925(20) + 7.644(25) + 4.297(40) - 37.354(45) = 1505.5$, and $t_{.025,25} = 2.060$. A 95% C.I for the true average energy content under these circumstances is $1505.5 \pm (2.060)(12.47) = 1505.5 \pm 25.69 = (1479.8, 1531.1)$. Because the interval is reasonably narrow, we would conclude that the mean energy content has been precisely estimated.

    e. A 95% prediction interval for the energy content of a waste sample having the specified characteristics is $1505.5 \pm (2.060)\sqrt{(31.48)^2 + (12.47)^2}$
$= 1505.5 \pm 69.75 = (1435.7, 1575.2)$.

49.

    a. $\hat{\mu}_{y \cdot 18.9, 43} = 21.967 = 96.8303$; Residual = 91 − 96.8303 = -5.8303.

    b. $H_0 : \beta_1 = \beta_2 = 0$; H_a : at least one $\beta_i \neq 0$; RR: $f \geq F_{.05,2,9} = 8.02$

$$f = \frac{R^2/k}{(1-R^2)/(n-k-1)} = \frac{.768/2}{(1-.768)/9} = 14.90.$$ Reject H_o. The model appears useful.

    c. $96.8303 \pm (2.262)(8.20) = (78.28, 115.38)$

    d. $96.8303 \pm (2.262)\sqrt{24.45^2 + 8.20^2} = (38.50, 155.16)$

51.

a. No, there is no pattern in the plots which would indicate that a transformation or the inclusion of other terms in the model would produce a substantially better fit.

b. $k = 5$, $n - (k+1) = 8$, so $H_0 : \beta_1 = ... = \beta_5 = 0$ will be rejected if

$$f \geq F_{.05,5,8} = 3.69; \quad f = \frac{(.759)/5}{(.241)/8} = 5.04 \geq 3.69, \text{ so we reject } H_0. \text{ At least}$$

one of the coefficients is not equal to zero.

c. When $x_1 = 8.0$ and $x_2 = 33.1$ the residual is $e = 2.71$ and the standardized residual is $e^* = .44$; since $e^* = e/(\text{sd of the residual})$, sd of residual $= e/e^* = 6.16$. Thus the estimated variance of $\hat{Y}$ is $(6.99)^2 - (6.16)^2 = 10.915$, so the estimated sd is 3.304. Since $\hat{y} = 24.29$ and $t_{.025,8} = 2.306$, the desired C.I. is

$$24.29 \pm 2.306(3.304) = (16.67, 31.91).$$

d. $F_{.05,3,8} = 4.07$, so $H_0 : \beta_3 = \beta_4 = \beta_5 = 0$ will be rejected if $f \geq 4.07$.

With $SSE_k = 8$, $s^2 = 390.88$, and $f = \frac{(894.95 - 390.88)/3}{(390.88)/8} = 3.44$, and since

3.44 is not ≥ 4.07, H_0 cannot be rejected and the quadratic terms should all be deleted. (n.b.: this is not a modification which would be suggested by a residual plot.

53. Some possible questions might be:

Is this model useful in predicting deposition of poly-aromatic hydrocarbons? A test of model utility gives us an F = 84.39, with a p-value of 0.000. Thus, the model is useful.

Is x_1 a significant predictor of y while holding x_2 constant? A test of $H_0 : \beta_1 = 0$ vs the two-tailed alternative gives us a t = 6.98 with a p-value of 0.000., so this predictor is significant.

A similar question, and solution for testing x_2 as a predictor yields a similar conclusion: With a p-value of 0.046, we would accept this predictor as significant if our significance level were anything larger than 0.046.

Section 13.5

55.

a. $\ln(Q) = Y = \ln(\alpha) + \beta\ln(a) + \gamma\ln(b) + \ln(\varepsilon) = \beta_0 + \beta_1 x_1 + \beta_2 x_2 + \varepsilon'$

where $x_1 = \ln(a), x_2 = \ln(b), \beta_0 = \ln(\alpha), \beta_1 = \beta, \beta_2 = \gamma$ and

$\varepsilon' = \ln(\varepsilon)$. Thus we transform to $(y, x_1, x_2) = (\ln(Q), \ln(a), \ln(b))$ (take the natural log of the values of each variable) and do a multiple linear regression. A computer analysis gave $\hat{\beta}_0 = 1.5652$, $\hat{\beta}_1 = .9450$, and $\hat{\beta}_2 = .1815$. For a = 10 and b = .01, $x_1 = \ln(10) = 2.3026$ and $x_2 = \ln(.01) = -4.6052$, from which $\hat{y} = 2.9053$ and $\hat{Q} = e^{2.9053} = 18.27$.

b. Again taking the natural log, $Y = \ln(Q) = \ln(\alpha) + \beta a + \gamma b + \ln(\varepsilon)$, so to fit this model it is necessary to take the natural log of each Q value (and not transform a or b) before using multiple regression analysis.

c. We simply exponentiate each endpoint: $(e^{.217}, e^{1.755}) = (1.24, 5.78)$.

57.

k	R^2	Adj. R^2	$C_k = \dfrac{SSE_k}{s^2} + 2(k+1) - n$
1	.676	.647	138.2
2	.979	.975	2.7
3	.9819	.976	3.2
4	.9824		4

Where $s^2 = 5.9825$

a. Clearly the model with k = 2 is recommended on all counts.

b. No. Forward selection would let x_4 enter first and would not delete it at the next stage.

59. The choice of a "best" model seems reasonably clear–cut. The model with 4 variables including all but the summerwood fiber variable would seem bests. R^2 is as large as any of the models, including the 5 variable model. R^2 adjusted is at its maximum and CP is at its minimum . As a second choice, one might consider the model with k = 3 which excludes the summerwood fiber and springwood % variables.

61. If multicollinearity were present, at least one of the four R^2 values would be very close to 1, which is not the case. Therefore, we conclude that multicollinearity is not a problem in this data.

63. We would need to investigate further the impact these two observations have on the equation. Removing observation #7 is reasonable, but removing #67 should be considered as well, before regressing again.

Supplementary Exercises

65.

  a.

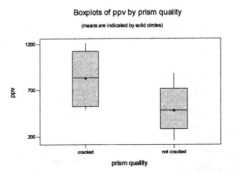

Boxplots of ppv by prism quality

(means are indicated by solid circles)

A two-sample t confidence interval, generated by Minitab:

```
Two sample T for ppv

prism qu      N      Mean      StDev    SE Mean
cracked       12     827       295         85
not cracke    18     483       234         55

95% CI for mu (cracked   ) - mu (not cracke): ( 132,   557)
```

276

b. The simple linear regression results in a significant model, r^2 is .577, but we have an extreme observation, with std resid = -4.11. Minitab output is below. Also run, but not included here was a model with an indicator for cracked/ not cracked, and for a model with the indicator and an interaction term. Neither improved the fit significantly.

```
The regression equation is
ratio = 1.00 -0.000018 ppv

Predictor         Coef        StDev           T         P
Constant       1.00161      0.00204      491.18     0.000
ppv         -0.00001827  0.00000295       -6.19     0.000

S = 0.004892    R-Sq = 57.7%      R-Sq(adj) = 56.2%

Analysis of Variance

Source            DF          SS          MS          F         P
Regression         1  0.00091571  0.00091571      38.26     0.000
Residual Error    28  0.00067016  0.00002393
Total             29  0.00158587

Unusual Observations
Obs        ppv      ratio        Fit  StDev Fit  Residual   St Resid
 29       1144   0.962000   0.980704   0.001786  -0.018704     -4.11R
R denotes an observation with a large standardized residual
```

67.

a. For a one-minute increase in the 1-mile walk time, we would expect the VO$_2$max to decrease by .0996, while keeping the other predictor variables fixed.

b. We would expect male to have an increase of .6566 in VO$_2$max over females, while keeping the other predictor variables fixed.

c. $\hat{y} = 3.5959 + .6566(1) + .0096(170) - .0996(11) - .0880(140) = 3.67$. The residual is $\hat{y} = (3.15 - 3.67) = -.52$.

d. $R^2 = 1 - \dfrac{SSE}{SST} = 1 - \dfrac{30.1033}{102.3922} = .706$, or 70.6% of the observed variations in VO$_2$max can be attributed to the model relationship.

e. $H_0 : \beta_1 = \beta_2 = \beta_3 = \beta_4 = 0$ will be rejected in favor of H_a : at least one among $\beta_1, ..., \beta_4 \neq 0$, if $f \geq F_{.05,4,15} = 8.25$. With

$$f = \frac{(.706)/4}{(1-.706)/15} = 9.005 \geq 8.25 \text{, so } H_o \text{ is rejected. It appears that the model}$$

specifies a useful relationship between VO_2max and at least one of the other predictors.

69.

a. Based on a scatter plot (below), a simple linear regression model would not be appropriate. Because of the slight, but obvious curvature, a quadratic model would probably be more appropriate.

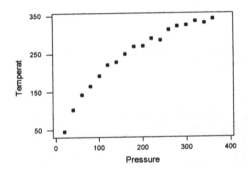

b. Using a quadratic model, a Minitab generated regression equation is $\hat{y} = 35.423 + 1.7191x - .0024753x^2$, and a point estimate of temperature when pressure is 200 is $\hat{y} = 280.23$. Minitab will also generate a 95% prediction interval of (256.25, 304.22). That is, we are confident that when pressure is 200 psi, a single value of temperature will be between 256.25 and 304.22 $^\circ F$.

71.

a. Using Minitab to generate the first order regression model, we test the model utility (to see if any of the predictors are useful), and with $f = 21.03$ and a p-value of .000, we determine that at least one of the predictors is useful in predicting palladium content. Looking at the individual predictors, the p-value associated with the pH predictor has value .169, which would indicate that this predictor is unimportant in the presence of the others.

b. Testing $H_0 : \beta_1 = ... = \beta_{20} = 0$ vs. H_a : at least one of the β_i's $\neq 0$. With calculated statistic $f = 6.29$, and p-value .002, this model is also useful at any reasonable significance level.

c. Testing $H_0 : \beta_6 = ... = \beta_{20} = 0$ vs. H_a : at least one of the listed β_i's $\neq 0$, the test statistic is $f = \dfrac{(SSE_l - SSE_k)/k-l}{(SSE_k)/n-k-1} = \dfrac{(716.10-290.27)/(20-5)}{290.27/(32-20-1)} = 1.07$. Using significance level .05, the rejection region would be $f \geq F_{.05,15,11} = 2.72$.

Since $1.07 < 2.72$, we fail to reject H_0 and conclude that all the quadratic and interaction terms should not be included in the model. They do not add enough information to make this model significantly better than the simple first order model.

73.

a. We wish to test $H_0 : \beta_1 = \beta_2 = 0$ vs. H_a : either β_1 or $\beta_2 \neq 0$. The test statistic is $f = \dfrac{R^2/k}{(1-R^2)/(n-k-1)}$, where k = 2 for the quadratic model. The rejection region is $f \geq F_{\alpha,k,n-k-1} = F_{.01,2,5} = 13.27$. $R^2 = 1 - \dfrac{.29}{202.88} = .9986$, giving f = 1783. No doubt about it, folks – the quadratic model is useful!

b. The relevant hypotheses are $H_0 : \beta_2 = 0$ vs. $H_a : \beta_2 \neq 0$. The test statistic value is $t = \dfrac{\hat{\beta}_2}{s_{\hat{\beta}_2}}$, and H_0 will be rejected at level .001 if either $t \geq 6.869$ or $t \leq -6.869$ (df = n – 3 = 5). Since $t = \dfrac{-.00163141}{.00003391} = -48.1 \leq -6.869$, H_0 is rejected. The quadratic predictor should be retained.

c. No. R^2 is extremely high for the quadratic model, so the marginal benefit of including the cubic predictor would be essentially nil – and a scatter plot doesn't show the type of curvature associated with a cubic model.

d. $t_{.025,5} = 2.571$, and $\hat{\beta}_0 + \hat{\beta}_1(100) + \hat{\beta}_2(100)^2 = 21.36$, so the C.I. is
$$21.36 \pm (2.571)(.1141) = 21.36 \pm .69 = (20.67, 22.05)$$

75.

a. $H_0 : \beta_1 = \beta_2 = 0$ will be rejected in favor of H_a : either β_1 or $\beta_2 \neq 0$ if
$$f = \frac{R^2/k}{(1-R^2)/(n-k-1)} \geq F_{\alpha,k,n-k-1} = F_{.01,2,7} = 9.55 . \ SST = \Sigma y^2 - \frac{(\Sigma y)}{n} = 264.5 \text{, so}$$
$$R^2 = 1 - \frac{26.98}{264.5} = .898 \text{, and } f = \frac{.898/2}{(.102)/7} = 30.8 . \text{ Because } 30.8 \geq 9.55 \ H_0 \text{ is}$$
rejected at significance level .01 and the quadratic model is judged useful.

b. The hypotheses are $H_0 : \beta_2 = 0$ vs. $H_a : \beta_2 \neq 0$. The test statistic value is
$$t = \frac{\hat{\beta}_2}{s_{\hat{\beta}_2}} = \frac{-2.3621}{.3073} = -7.69 \text{, and } t_{.0005,7} = 5.408 \text{, so } H_0 \text{ is rejected at level}$$
.001 and p-value < .001. The quadratic predictor should not be eliminated.

c. $x = 1$ here, and $\hat{\mu}_{Y\cdot 1} = \hat{\beta}_0 + \hat{\beta}_1(1) + \hat{\beta}_2(1)^2 = 45.96 . \ t_{.025,7} = 1.895$, giving the C.I. $45.96 \pm (1.895)(1.031) = (44.01, 47.91)$.

77.

a. Estimate $= \hat{\beta}_0 + \hat{\beta}_1(15) + \hat{\beta}_2(3.5)^2 = 180 + (1)(15) + (10.5)(3.5) = 231.75$

b. $R^2 = 1 - \dfrac{117.4}{1210.30} = .903$

c. $H_0 : \beta_1 = \beta_2 = 0$ vs. H_a : either β_1 or $\beta_2 \neq 0$ (or both). $f = \dfrac{.903/2}{.097/9} = 41.9$,

which greatly exceeds $F_{.01,2,9}$ so there appears to be a useful linear relationship.

d. $s^2 = \dfrac{117.40}{12-3} = 13.044$, $\sqrt{s^2 + (est.st.dev)^2} = 3.806$, $t_{.025,9} = 2.262$. The P.I.

is $229.5 \pm (2.262)(3.806) = (220.9, 238.1)$

79. There are obviously several reasonable choices in each case. In **a**, the model with 6 carriers is a defensible choice on all three grounds, as are those with 7 and 8 carriers. The models with 7, 8, or 9 carriers in **b** merit serious consideration. These models merit consideration because R_k^2, MSE_k, and CK meet the variable selection criteria given in Section 13.5.

81.

a. The relevant hypotheses are $H_0 : \beta_1 = \ldots = \beta_5 = 0$ vs. H_a: at least one among $\beta_1, \ldots, \beta_5$ is not 0. $F_{.05,5,111} = 2.29$ and $f = \dfrac{(.827)/(5)}{(.173)/(111)} = 106.1$. Because $106.1 \geq 2.29$, H_0 is rejected in favor of the conclusion that there is a useful linear relationship between Y and at least one of the predictors.

b. $t_{.05,111} = 1.66$, so the C.I. is
$$.041 \pm (1.66)(.016) = .041 \pm .027 = (.014, .068).$$ β_1 is the expected change in mortality rate associated with a one-unit increase in the particle reading when the other four predictors are held fixed; we cab be 90% confident that $.014 < \beta_1 < .068$.

c. $H_0 : \beta_4 = 0$ will be rejected in favor of $H_a : \beta_4 \neq 0$ if $t = \dfrac{\hat{\beta}_4}{s_{\hat{\beta}_4}}$ is either ≥ 2.62 or ≤ -2.62. $t = \dfrac{.014}{.007} = 5.9 \geq 2.62$, so H_0 is rejected and this predictor is judged important.

d. $\hat{y} = 19.607 + .041(166) + .071(60) + .001(788) + .041(68) + .687(.95) = 99.514$ and the corresponding residual is $103 - 99.514 = 3.486$.

CHAPTER 14

Section 14.1

1.

   a. We reject H_o if the calculated χ^2 value is greater than or equal to the tabled value of $\chi^2_{\alpha, k-1}$ from Table A.7. Since $12.25 \geq \chi^2_{.05,4} = 9.488$, we would reject H_o.

   b. Since 8.54 is not $\geq \chi^2_{.01,3} = 11.344$, we would fail to reject H_o.

   c. Since 4.36 is not $\geq \chi^2_{.10,2} = 4.605$, we would fail to reject H_o.

   d. Since 10.20 is not $\geq \chi^2_{.01,5} = 15.085$, we would fail to reject H_o.

3.   Using the number 1 for business, 2 for engineering, 3 for social science, and 4 for agriculture, let p_i = the true proportion of all clients from discipline i. If the Statistics department's expectations are correct, then the relevant null hypothesis is $H_o : p_1 = .40, p_2 = .30, p_3 = .20, p_4 = .10$, versus H_a : The Statistics department's expectations are not correct. With d.f $= k - 1 = 4 - 1 = 3$, we reject H_o if $\chi^2 \geq \chi^2_{.05,3} = 7.815$. Using the proportions in H_o, the expected number of clients are :

Client's Discipline	Expected Number
Business	(120)(.40) = 48
Engineering	(120)(.30) = 36
Social Science	(120)(.20) = 24
Agriculture	(120)(.10) = 12

Since all the expected counts are at least 5, the chi-squared test can be used. The value of the test statistic is

$$\chi^2 = \sum_{i=1}^{k} \frac{(n_i - np_i)^2}{np_i} = \sum_{allcells} \frac{(observed - expected)^2}{expected}$$

$$= \left[\frac{(52 - 48)^2}{48} + \frac{(38 - 36)^2}{36} + \frac{(21 - 24)^2}{24} + \frac{(9 - 12)^2}{12} \right] = 1.57 \text{, which is not}$$

≥ 7.815, so we fail to reject H_o. (Alternatively, p-value $= P(\chi^2 \geq 1.57)$ which is $> .10$, and since the p-value is not $< .05$, we reject H_o). Thus we have no evidence to suggest that the statistics department's expectations are incorrect.

5.   We will reject H_o if the p-value $< .10$. The observed values, expected values, and corresponding χ^2 terms are :

Obs	4	15	23	25	38	21	32	14	10	8
Exp	6.67	13.33	20	26.67	33.33	33.33	26.67	20	13.33	6.67
χ^2	1.069	.209	.450	.105	.654	.163	1.065	1.800	.832	.265

$\chi^2 = 1.069 + ... + .265 = 6.612$. With d.f. $= 10 - 1 = 9$, our χ^2 value of 6.612 is less than $\chi^2_{.10,9} = 14.684$, so the p-value $> .10$, which is not $< .10$, so we cannot reject H_o. There is no evidence that the data is not consistent with the previously determined proportions.

Chapter 14: The Analysis of Categorical Data

7.   We test $H_o : p_1 = p_2 = p_3 = p_4 = .25$ vs. H_a :at least one proportion $\neq .25$, and d.f. = 3. We will reject H_o if the p-value $< .01$.

Cell	1	2	3	4
Observed	328	334	372	327
Expected	340.25	340.25	340.25	34.025
χ^2 term	.4410	.1148	2.9627	.5160

$\chi^2 = 4.0345$, and with 3 d.f., p-value $> .10$, so we fail to reject H_o. The data fails to indicate a seasonal relationship with incidence of violent crime.

9.

a.   Denoting the 5 intervals by $[0, c_1), [c_1, c_2), ..., [c_4, \infty)$, we wish c_1 for which

$$.2 = P(0 \leq X \leq c_1) = \int_0^{c_1} e^{-x} dx = 1 - e^{-c_1}$$, so $c_1 = -\ln(.8) = .2231$. Then

$$.2 = P(c_1 \leq X \leq c_2) \Rightarrow .4 = P(0 \leq X_1 \leq c_2) = 1 - e^{-c_2}$$, so $c_2 = -\ln(.6) =$ .5108. Similarly, $c_3 = -\ln(.4) = .0163$ and $c_4 = -\ln(.2) = 1.6094$. the resulting intervals are $[0, .2231), [.2231, .5108), [.5108, .9163), [.9163, 1.6094)$, and $[1.6094, \infty)$.

b.   Each expected cell count is $40(.2) = 8$, and the observed cell counts are 6, 8, 10,

7, and 9, so $\chi^2 = \left[\dfrac{(6-8)^2}{8} + ... + \dfrac{(9-8)^2}{8} \right] = 1.25$. Because 1.25 is not

$\geq \chi^2_{.10,4} = 7.779$, even at level .10 H_o cannot be rejected; the data is quite consistent with the specified exponential distribution.

11.

a.   The six intervals must be symmetric about 0, so denote the 4th, 5th and 6th intervals by [0, a0, [a, b), [b, ∞). a must be such that $\Phi(a) = .6667(\frac{1}{2} + \frac{1}{6})$, which from Table A.3 gives $a \approx .43$. Similarly $\Phi(b) = .8333$ implies $b \approx .97$, so the six intervals are $(-\infty, -.97), [-.97, -.43), [-.43, 0), [0, .43), [.43, .97)$, and $[.97, \infty)$.

b.   The six intervals are symmetric about the mean of .5. From **a**, the fourth interval should extend from the mean to .43 standard deviations above the mean, i.e., from .5 to .5 + .43(.002), which gives [.5, .50086). Thus the third interval is [.5 - .00086, .5) = [.49914, .5). Similarly, the upper endpoint of the fifth interval is .5 + .97(.002) = .50194, and the lower endpoint of the second interval is .5 - .00194 = .49806. The resulting intervals are $(-\infty, .49806), [.49806, .49914), [.49914, .5), [.5, .50086), [.50086, .50194)$, and $[.50194, \infty)$.

c.   Each expected count is $45(\frac{1}{6}) = 7.5$, and the observed counts are 13, 6, 6, 8, 7, and 5, so $\chi^2 = 5.53$. With 5 d.f., the p-value > .10, so we would fail to reject H_o at any of the usual levels of significance. There is no evidence to suggest that the bolt diameters are not normally distributed.

Section 14.2

13.   According to the stated model, the three cell probabilities are $(1-p)^2$, $2p(1-p)$, and p^2, so we wish the value of p which maximizes $(1-p)^{2n_1}[2p(1-p)]^{n_2}p^{2n_3}$. Proceeding as in example 14.6 gives $\hat{p} = \dfrac{n_2 + 2n_3}{2n} = \dfrac{234}{2776} = .0843$. The estimated expected cell counts are then $n(1-\hat{p})^2 = 1163.85$, $n[2\hat{p}(1-\hat{p})]^2 = 214.29$, $n\hat{p}^2 = 9.86$. This gives

$$\chi^2 = \left[\frac{(1212-1163.85)^2}{1163.85} + \frac{(118-214.29)^2}{214.29} + \frac{(58-9.86)^2}{9.86} \right] = 280.3.$$

According to (14.15), H_o will be rejected if $\chi^2 \geq \chi^2_{\alpha,2}$, and since $\chi^2_{.01,2} = 9.210$, H_o is soundly rejected; the stated model is strongly contradicted by the data.

Chapter 14: The Analysis of Categorical Data

15. The part of the likelihood involving θ is

$$\left[(1-\theta)^4\right]^{n_1} \cdot \left[\theta(1-\theta)^3\right]^{n_2} \cdot \left[\theta^2(1-\theta)^2\right]^{n_3} \cdot$$

$$\left[\theta^3(1-\theta)\right]^{n_4} \cdot \left[\theta^4\right]^{n_5} = \theta^{n_2+2n_3+3n_4+4n_5}(1-\theta)^{4n_1+3n_2+2n_3+n_4} = \theta^{233}(1-\theta)^{367}, \text{ so}$$

$\ln(likelihood) = 233\ln\theta + 367\ln(1-\theta)$. Differentiating and equating to 0

yields $\hat\theta = \dfrac{233}{600} = .3883$, and $(1-\hat\theta) = .6117$ [note that the exponent on θ is

simply the total # of successes (defectives here) in the n = 4(150) = 600 trials.]
Substituting this θ' into the formula for p_i yields estimated cell probabilities .1400,
.3555, .3385, .1433, and .0227. Multiplication by 150 yields the estimated expected
cell counts are 21.00, 53.33, 50.78, 21.50, and 3.41. the last estimated expected cell
count is less than 5, so we combine the last two categories into a single one (≥ 3
defectives), yielding estimated counts 21.00, 53.33, 50.78, 24.91, observed counts 26,
51, 47, 26, and $\chi^2 = 1.62$. With d.f. = 4 – 1 – 1 = 2, since

$1.62 < \chi^2_{.10,2} = 4.605$, the p-value > .10, and we do not reject H_o. The data
suggests that the stated binomial distribution is plausible.

17. $\hat\lambda = \dfrac{380}{120} = 3.167$, so $\hat p = e^{-3.167}\dfrac{(3.167)^x}{x!}$.

x	0	1	2	3	4	5	6	≥ 7
$\hat p$	.0421	.1334	.2113	.2230	.1766	.1119	.0590	.0427
$n\hat p$	5.05	16.00	25.36	26.76	21.19	13.43	7.08	5.12
obs	24	16	16	18	15	9	6	16

The resulting value of $\chi^2 = 103.98$, and when compared to $\chi^2_{.01,7} = 18.474$, it is
obvious that the Poisson model fits very poorly.

19. With $A = 2n_1 + n_4 + n_5$, $B = 2n_2 + n_4 + n_6$, and $C = 2n_3 + n_5 + n_6$, the likelihood is proportional to $\theta_1^A \theta_2^B (1 - \theta_1 - \theta_2)^C$, where $A + B + C = 2n$. Taking the natural log and equating both $\dfrac{\partial}{\partial \theta_1}$ and $\dfrac{\partial}{\partial \theta_2}$ to zero gives $\dfrac{A}{\theta_1} = \dfrac{C}{1 - \theta_1 - \theta_2}$ and

$\dfrac{B}{\theta_2} = \dfrac{C}{1 - \theta_1 - \theta_2}$, whence $\theta_2 = \dfrac{B\theta_1}{A}$. Substituting this into the first equation

gives $\theta_1 = \dfrac{A}{A + B + C}$, and then $\theta_2 = \dfrac{B}{A + B + C}$. Thus $\hat{\theta}_1 = \dfrac{2n_1 + n_4 + n_5}{2n}$,

$\hat{\theta}_2 = \dfrac{2n_2 + n_4 + n_6}{2n}$, and $\left(1 - \hat{\theta}_1 - \hat{\theta}_2\right) = \dfrac{2n_3 + n_5 + n_6}{2n}$. Substituting the

observed n_i's yields $\hat{\theta}_1 = \dfrac{2(49) + 20 + 53}{400} = .4275$, $\hat{\theta}_2 = \dfrac{110}{400} = .2750$, and

$\left(1 - \hat{\theta}_1 - \hat{\theta}_2\right) = .2975$, from which $\hat{p}_1 = (.4275)^2 = .183$,

$\hat{p}_2 = .076$, $\hat{p}_3 = .089$, $\hat{p}_4 = 2(.4275)(.275) = .235$, $\hat{p}_5 = .254$,

$\hat{p}_6 = .164$.

Category	1	2	3	4	5	6
np	36.6	15.2	17.8	47.0	50.8	32.8
observed	49	26	14	20	53	38

This gives $\chi^2 = 29.1$. With $\chi^2_{.01,6-1-2} = \chi^2_{.01,3} = 11.344$, and

$\chi^2_{.01,6-1} = \chi^2_{.01,5} = 15.085$, according to (14.15) H_o must be rejected since $29.1 \geq 15.085$.

21. The Ryan-Joiner test p-value is larger than .10, so we conclude that the null hypothesis of normality cannot be rejected. This data could reasonably have come from a normal population. This means that it would be legitimate to use a one-sample t test to test hypotheses about the true average ratio.

23. Minitab gives r = .967, though the hand calculated value may be slightly different because when there are ties among the $x_{(i)}$'s, Minitab uses the same y_i for each $x_{(i)}$ in a group of tied values. $C_{10} = .9707$, and $c_{.05} = 9639$, so $.05 < $ p-value $< .10$. At the 5% significance level, one would have to consider population normality plausible.

Section 14.3

25. Let P_{ij} = the proportion of white clover in area of type i which has a type j mark (i = 1, 2; j = 1, 2, 3, 4, 5). The hypothesis H_o: $p_{1j} = p_{2j}$ for j = 1, ..., 5 will be rejected at level .01 if $\chi^2 \geq \chi^2_{.01,(2-1)(5-1)} = \chi^2_{.01,4} = 13.277$.

$\hat{E}_{ij}$	1	2	3	4	5		
1	449.66	7.32	17.58	8.79	242.65	726	$\chi^2 = 23.18$
2	471.34	7.68	18.42	9.21	254.35	761	
	921	15	36	18	497	1487	

Since $23.18 \geq 13.277$, H_o is rejected.

27. With i = 1 identified with men and i = 2 identified with women, and j = 1, 2, 3 denoting the 3 categories L>R, L=R, L<R, we wish to test H_o: $p_{1j} = p_{2j}$ for j = 1, 2, 3 vs. H_a: p_{1j} not equal to p_{2j} for at least one j. The estimated cell counts for men are 17.95, 8.82, and 13.23 and for women are 39.05, 19.18, 28.77, resulting in $\chi^2 = 44.98$. With (2 − 1)(3 − 1) = 2 degrees of freedom, since $44.98 > \chi^2_{.005,2} = 10.597$, p-value < .005, which strongly suggests that H_o should be rejected.

29. H_o: $p_{1j} = ... = p_{6j}$ for j = 1, 2, 3 is the hypothesis of interest, where p_{ij} is the proportion of the jth sex combination resulting from the ith genotype. H_o will be rejected at level .10 if $\chi^2 \geq \chi^2_{.10,10} = 15.987$.

$\hat{E}_{ij}$	1	2	3		χ^2	1	2	3
1	35.8	83.1	35.1	154		.02	.12	.44
2	39.5	91.8	38.7	170		.06	.66	1.01
3	35.1	81.5	34.4	151		.13	.37	.34
4	9.8	22.7	9.6	42		.32	.49	.26
5	5.1	11.9	5.0	22		.00	.06	.19
6	26.7	62.1	26.2	115		.40	.14	1.47
	152	353	149	654				6.46

(carrying 2 decimal places in $\hat{E}_{ij}$ yields $\chi^2 = 6.49$). Since 6.46 < 15.987, H_o cannot be rejected at level .10.

31. With I denoting the Ith type of car ($I = 1, 2, 3, 4$) and j the jth category of commuting distance, H_o: $p_{ij} = p_{i.} \ p_{.j}$ (type of car and commuting distance are independent) will be rejected at level .05 if $\chi^2 \geq \chi^2_{.05,6} = 12.592$.

$\hat{E}_{ij}$	1	2	3	
1	10.19	26.21	15.60	52
2	11.96	30.74	18.30	61
3	19.40	49.90	29.70	99
4	7.45	19.15	11.40	38
	49	126	75	250

$\chi^2 = 14.15 \geq 12.592$, so the independence hypothesis H_o is rejected at level .05 (but not at level .025!)

33. $$\chi^2 = \Sigma\Sigma\frac{\left(N_{ij} - \hat{E}_{ij}\right)^2}{\hat{E}_{ij}} = \Sigma\Sigma\frac{N_{ij}^2 - 2\hat{E}_{ij}N_{ij} + \hat{E}_{ij}^2}{\hat{E}_{ij}} = \frac{\Sigma\Sigma N_{ij}^2}{\hat{E}_{ij}} - 2\Sigma\Sigma N_{ij} + \Sigma\Sigma\hat{E}_{ij},$$

but $\Sigma\Sigma\hat{E}_{ij} = \Sigma\Sigma N_{ij} = n$, so $\chi^2 = \Sigma\Sigma\frac{N_{ij}^2}{\hat{E}_{ij}} - n$. This formula is computationally

efficient because there is only one subtraction to be performed, which can be done as the last step in the calculation.

35. With p_{ij} denoting the common value of p_{ij1}, p_{ij2}, p_{ij3}, p_{ij4} (under H_o), $\hat{p}_{ij} = \dfrac{N_{ij}}{n}$ and

$\hat{E}_{ijk} = \dfrac{n_k N_{ij}}{n}$. With four different tables (one for each region), there are $8 + 8 + 8$

$+ 8 = 32$ freely determined cell counts. Under H_o, p_{11}, ..., p_{33} must be estimated but $\Sigma\Sigma p_{ij} = 1$ so only 8 independent parameters are estimated, giving χ^2 df $= 32 - 8 = 24$.

Chapter 14: The Analysis of Categorical Data

Supplementary Exercises

37. There are 3 categories here – firstborn, middleborn, (2nd or 3rd born), and lastborn. With p_1, p_2, and p_3 denoting the category probabilities, we wish to test H_o: $p_1 = .25$, $p_2 = .50$ ($p_2 = P(2^{nd}$ or 3rd born) $= .25 + .25 = .50$), $p_3 = .25$. H_o will be rejected at significance level .05 if $\chi^2 \geq \chi^2_{.05,2} = 5.992$. The expected counts are $(31)(.25) = 7.75$, $(31)(.50) = 15.5$, and 7.75, so
$$\chi^2 = \frac{(12-7.75)^2}{7.75} + \frac{(11-15.5)^2}{15.5} + \frac{(8-7.75)^2}{7.75} = 3.65.$$ Because $3.65 < 5.992$, H_o is not rejected. The hypothesis of equiprobable birth order appears quite plausible.

39. H_o: gender and years of experience are independent; H_a: gender and years of experience are not independent. Df $= 4$, and we reject H_o if $\chi^2 \geq \chi^2_{.01,4} = 13.277$.

Gender		Years of Experience				
		$1-3$	$4-6$	$7-9$	$10-12$	$13+$
Male Observed		202	369	482	361	811
Expected		285.56	409.83	475.94	347.04	706.63
$\frac{(O-E)^2}{E}$		24.451	4.068	.077	.562	15.415
Female Observed		230	251	238	164	258
Expected		146.44	210.17	244.06	177.96	362.37
$\frac{(O-E)^2}{E}$		47.680	7.932	.151	1.095	30.061

$\chi^2 = \Sigma \frac{(O-E)^2}{E} = 131.492$. Reject H_o. The two variables do not appear to be independent. In particular, women have higher than expected counts in the beginning category ($1-3$ years) and lower than expected counts in the more experienced category ($13+$ years).

41.   The null hypothesis H_o: $p_{ij} = p_{i.} \, p_{.j}$ states that level of parental use and level of student use are independent in the population of interest. The test is based on $(3-1)(3-1) = 4$ df.

Estimated Expected

119.3	57.6	58.1	235
82.8	33.9	40.3	163
23.9	11.5	11.6	47
226	109	110	445

The calculated value of $\chi^2 = 22.4$. Since $22.4 > \chi^2_{.005,4} = 14.860$, p-value < .005, so H_o should be rejected at any significance level greater than .005. Parental and student use level do not appear to be independent.

43.   This is a test of homogeneity: H_o: $p_{1j} = p_{2j} = p_{3j}$ for $j = 1, 2, 3, 4, 5$. The given SPSS output reports the calculated $\chi^2 = 70.64156$ and accompanying p-value (significance) of .0000. We reject H_o at any significance level. The data strongly supports that there are differences in perception of odors among the three areas.

45. $(n_1 - np_{10})^2 = (np_{10} - n_1)^2 = (n - n_1 - n(1 - p_{10}))^2 = (n_2 - np_{20})^2$. Therefore

$$\chi^2 = \frac{(n_1 - np_{10})^2}{np_{10}} + \frac{(n_2 - np_{20})^2}{np_{20}} = \frac{(n_1 - np_{10})^2}{n_2}\left(\frac{n}{p_{10}} + \frac{n}{p_{20}}\right)$$

$$= \left(\frac{n_1}{n} - p_{10}\right)^2 \cdot \left(\frac{n}{p_{10}p_{20}}\right) = \frac{(\hat{p}_1 - p_{10})^2}{p_{10}p_{20}/n} = z^2.$$

47.

a. Our hypotheses are H_0: no difference in proportion of concussions among the three groups. Vs H_a: there is a difference …

Observed	Concussion	No Concussion	Total
Soccer	45	46	91
Non Soccer	28	68	96
Control	8	45	53
Total	81	159	240

Expected	Concussion	No Concussion	Total
Soccer	30.7125	60.2875	91
Non Soccer	32.4	63.6	96
Control	17.8875	37.1125	53
Total	81	159	240

$$\chi^2 = \frac{(45-30.7125)^2}{30.7125} + \frac{(46-60.2875)^2}{60.2875} + \frac{(28-32.4)^2}{32.4} + \frac{(68-63.6)^2}{63.6}$$
$$+ \frac{(8-17.8875)^2}{17.8875} + \frac{(45-37.1125)^2}{37.1125} = 19.1842.$$ The df for this test is (I –

1)(J – 1) = 2, so we reject H_o if $\chi^2 > \chi^2_{.05,2} = 5.99$. $19.1842 > 5.99$, so we reject H_0. There is a difference in the proportion of concussions based on whether a person plays soccer.

b. We are testing the hypothesis H_0: $\rho = 0$ vs H_a: $\rho \neq 0$. The test statistic is

$$t = \frac{r\sqrt{n-2}}{\sqrt{1-r^2}} = \frac{-.22\sqrt{89}}{\sqrt{1-.22^2}} = -2.13.$$ At significance level $\alpha = .01$, we would

fail to reject and conclude that there is no evidence of non-zero correlation in the population. If we were willing to accept a higher significance level, our decision could change. At best, there is evidence of only weak correlation.

c. We will test to see if the average score on a controlled word association test is the same for soccer and non-soccer athletes. $H_0: \mu_1 = \mu_2$ vs $H_a: \mu_1 \neq \mu_2$. We'll use

test statistic $t = \dfrac{(\bar{x}_1 - \bar{x}_2)}{\sqrt{\dfrac{s_1^2}{m} + \dfrac{s_2^2}{n}}}$. With $\dfrac{s_1^2}{m} = 3.206$ and $\dfrac{s_2^2}{n} = 1.854$,

$t = \dfrac{(37.50 - 39.63)}{\sqrt{3.206 + 1.854}} = -.95$. The df $= \dfrac{(3.206 + 1.854)^2}{\dfrac{3.206^2}{25} + \dfrac{1.854^2}{55}} \approx 56$. The p-

value will be $> .10$, so we do not reject H_0 and conclude that there is no difference in the average score on the test for the two groups of athletes.

d. Our hypotheses for ANOVA are H_0: all means are equal vs H_a: not all means are equal. The test statistic is $f = \dfrac{MSTr}{MSE}$.

$SSTr = 91(.30 - .35)^2 + 96(.49 - .35)^2 + 53(.19 - .35)^2 = 3.4659$

$MSTr = \dfrac{3.4659}{2} = 1.73295$

$SSE = 90(.67)^2 + 95(.87)^2 + 52(.48)^2 = 124.2873$ and

$MSE = \dfrac{124.2873}{237} = .5244$. Now, $f = \dfrac{1.73295}{.5244} = 3.30$. Using df 2,200 from table A.9, the p value is between .01 and .05. At significance level .05, we reject the null hypothesis. There is sufficient evidence to conclude that there is a difference in the average number of prior non-soccer concussions between the three groups.

CHAPTER 15

Section 15.1

1. We test $H_0 : \mu = 100$ vs. $H_a : \mu \neq 100$. The test statistic is s_+ = sum of the ranks associated with the positive values of $(x_i - 100)$, and we reject H_0 at significance level .05 if $s_+ \geq 64$. (from Table A.13, n = 12, with $\alpha / 2 = .026$, which is close to the desired value of .025), or if

$$s_+ \leq \frac{12(13)}{2} - 64 = 78 - 64 = 14.$$

x_i	$(x_i - 100)$	ranks
105.6	5.6	7*
90.9	-9.1	12
91.2	-8.8	11
96.9	-3.1	3
96.5	-3.5	5
91.3	-8.7	10
100.1	0.1	1*
105	5	6*
99.6	-0.4	2
107.7	7.7	9*
103.3	3.3	4*
92.4	-7.6	8

$S_+ = 27$, and since 27 is neither ≥ 64 nor ≤ 14, we do not reject H_0. There is not enough evidence to suggest that the mean is something other than 100.

3. We test $H_0 : \mu = 7.39$ vs. $H_a : \mu \neq 7.39$, so a two tailed test is appropriate. With n = 14 and $\alpha / 2 = .025$, Table A.13 indicates that H_0 should be rejected if either $s_+ \geq 84 or \leq 21$. The $(x_i - 7.39)$'s are -.37, -.04, -.05, -.22, -.11, .38, -.30, -.17, .06, -.44, .01, -.29, -.07, and -.25, from which the ranks of the three positive differences are 1, 4, and 13. Since $s_+ = 18 \leq 21$, H_0 is rejected at level .05.

5. The data is paired, and we wish to test $H_0 : \mu_D = 0$ vs. $H_a : \mu_D \neq 0$. With n = 12 and $\alpha = .05$, H_0 should be rejected if either $s_+ \geq 64$ or if $s_+ \leq 14$.

d_i	-.3	2.8	3.9	.6	1.2	-1.1	2.9	1.8	.5	2.3	.9	2.5
rank	1	10*	12*	3*	6*	5	11*	7*	2*	8*	4*	9*

$s_+ = 72$, and $72 \geq 64$, so H_0 is rejected at level .05. In fact for $\alpha = .01$, the critical value is c = 71, so even at level .01 H_0 would be rejected.

7. $H_0 : \mu_D = .20$ vs. $H_a : \mu_D > .20$, where $\mu_D = \mu_{outdoor} - \mu_{indoor}$. $\alpha = .05$, and because n = 33, we can use the large sample test. The test statistic is

$$Z = \frac{S_+ - \frac{n(n+1)}{4}}{\sqrt{\frac{n(n+1)(2n+1)}{24}}}, \text{ and we reject } H_0 \text{ if } z \geq 1.96.$$

d_i	$d_i - .2$	rank	d_i	$d_i - .2$	rank	d_i	$d_i - .2$	rank
0.22	0.02	2	0.15	-0.05	5.5	0.63	0.43	23
0.01	-0.19	17	1.37	1.17	32	0.23	0.03	4
0.38	0.18	16	0.48	0.28	21	0.96	0.76	31
0.42	0.22	19	0.11	-0.09	8	0.2	0	1
0.85	0.65	29	0.03	-0.17	15	-0.02	-0.22	18
0.23	0.03	3	0.83	0.63	28	0.03	-0.17	14
0.36	0.16	13	1.39	1.19	33	0.87	0.67	30
0.7	0.5	26	0.68	0.48	25	0.3	0.1	9.5
0.71	0.51	27	0.3	0.1	9.5	0.31	0.11	11
0.13	-0.07	7	-0.11	-0.31	22	0.45	0.25	20
0.15	-0.05	5.5	0.31	0.11	12	-0.26	-0.46	24

$s_+ = 434$, so $z = \dfrac{424 - 280.5}{\sqrt{3132.25}} = \dfrac{143.5}{55.9665} = 2.56$. Since $2.56 \geq 1.96$, we reject H_0 at significance level .05.

9.

r_1	1	1	1	1	1	1	2	2	2	2	2	2
r_2	2	2	3	3	4	4	1	1	3	3	4	4
r_3	3	4	2	4	2	3	3	4	1	4	1	3
r_4	4	3	4	2	3	2	4	3	4	1	3	1
D	0	2	2	6	6	8	2	4	6	12	10	14

r_1	3	3	3	3	3	3	4	4	4	4	4	4
r_2	1	1	2	2	4	4	1	1	2	2	3	3
r_3	2	4	1	4	1	2	2	3	1	3	1	2
r_4	4	2	4	1	2	1	3	2	3	1	2	1
D	6	10	8	14	16	18	12	14	14	18	18	20

When H_o is true, each of the above 24 rank sequences is equally likely, which yields the distribution of D when H_o is true as described in the answer section (e.g., P(D = 2) = P(1243 or 1324 or 2134) = 3/24). Then c = 0 yields $\alpha = \frac{1}{24} = .042$ while c = 2 implies $\alpha = \frac{4}{24} = .167$.

Section 15.2

9. With X identified with pine (corresponding to the smaller sample size) and Y with oak, we wish to test $H_0 : \mu_1 - \mu_2 = 0$ vs. $H_a : \mu_1 - \mu_2 \neq 0$. From Table A.14 with m = 6 and n = 8, H_o is rejected in favor of H_a at level .05 if either $w \geq 61$ or if $w \leq 90 - 61 = 29$ (the actual α is 2(.021) = .042). The X ranks are 3 (for .73), 4 (for .98), 5 (for 1.20), 7 (for 1.33), 8 (for 1.40), and 10 (for 1.52), so w = 37. Since 37 is neither ≥ 61 nor ≤ 29, H_o cannot be rejected.

Chapter 15: Distribution-Free Procedures

11. The hypotheses of interest are $H_0 : \mu_1 - \mu_2 = 1$ vs. $H_a : \mu_1 - \mu_2 > 1$, where 1(X) refers to the original process and 2 (Y) to the new process. Thus 1 must be subtracted from each x_1 before pooling and ranking. At level .05, H_o should be rejected in favor of H_a if $w \geq 84$.

x − 1	3.5	4.1	4.4	4.7	5.3	5.6	7.5	7.6
rank	1	4	5	6	8	10	15	16
y	3.8	4.0	4.9	5.5	5.7	5.8	6.0	7.0
rank	2	3	7	9	11	12	13	14

Since w = 65, H_o is not rejected.

13. Here m = n = 10 > 8, so we use the large-sample test statistic from p. 663.
$H_0 : \mu_1 - \mu_2 = 0$ will be rejected at level .01 in favor of $H_a : \mu_1 - \mu_2 \neq 0$ if either $z \geq 2.58$ or $z \leq -2.58$. Identifying X with orange juice, the X ranks are 7, 8, 9, 10, 11, 16, 17, 18, 19, and 20, so w = 135. With $\dfrac{m(m+n+1)}{2} = 105$ and

$\sqrt{\dfrac{mn(m+n+1)}{12}} = \sqrt{175} = 13.22$, $z = \dfrac{135-105}{13.22} = 2.27$. Because 2.27 is neither ≥ 2.58 nor ≤ -2.58, H_o is not rejected.
$p - value \approx 2(1 - \Phi(2.27)) = .0232$.

15. Let μ_1 and μ_2 denote true average cotanine levels in unexposed and exposed infants, respectively. The hypotheses of interest are $H_0 : \mu_1 - \mu_2 = -25$ vs. $H_a : \mu_1 - \mu_2 < -25$. With m = 7, n = 8, H_o will be rejected at level .05 if $w \leq 7(7+8+1)-71 = 41$. Before ranking, -25 is subtracted from each x_1 (i.e. 25 is added to each), giving 33, 36, 37, 39, 45, 68, and 136. The corresponding ranks in the combined set of 15 observations are 1, 3, 4, 5, 6, 8, and 12, from which w = 1 + 3 + ... + 12 = 39. Because $39 \leq 41$, H_o is rejected. The true average level for exposed infants appears to exceed that for unexposed infants by more than 25 (note that H_o would not be rejected using level .01).

Section 15.3

17. $n = 8$, so from Table A.15, a 95% C.I. (actually 94.5%) has the form
$\left(\overline{x}_{(36-32+1)}, \overline{x}_{(32)} \right) = \left(\overline{x}_{(5)}, \overline{x}_{(32)} \right)$. It is easily verified that the 5 smallest pairwise

averages are $\dfrac{5.0 + 5.0}{2} = 5.00$, $\dfrac{5.0 + 11.8}{2} = 8.40$, $\dfrac{5.0 + 12.2}{2} = 8.60$,

$\dfrac{5.0 + 17.0}{2} = 11.00$, and $\dfrac{5.0 + 17.3}{2} = 11.15$ (the smallest average not involving

5.0 is $\overline{x}_{(6)} = \dfrac{11.8 + 11.8}{2} = 11.8$), and the 5 largest averages are 30.6, 26.0, 24.7,

23.95, and 23.80, so the confidence interval is (11.15, 23.80).

19. The ordered d_i's are $-13, -12, -11, -7, -6$; with $n = 5$ and $\dfrac{n(n+1)}{2} = 15$, Table A.15

shows the 94% C.I. as (since $c = 1$) $\left(\overline{d}_{(1)}, \overline{d}_{(15)} \right)$. The smallest average is clearly

$\dfrac{-13 - 13}{2} = -13$ while the largest is $\dfrac{-6 - 6}{2} = -6$, so the C.I. is $(-13, \ -6)$.

21. $m = n = 5$ and from Table A.16, $c = 21$ and the 90% (actually 90.5%) interval is
$\left(d_{ij(5)}, d_{ij(21)} \right)$. The five smallest $x_i - y_j$ differences are $-18, -2, 3, 4, 16$ while the
five largest differences are 136, 123, 120, 107, 86 (construct a table like Table 15.5),
so the desired interval is $(16, 86)$.

Section 15.4

23. Below we record in parentheses beside each observation the rank of that observation in the combined sample.

1:	5.8(3)	6.1(5)	6.4(6)	6.5(7)	7.7(10)	$r_1 = 31$
2:	7.1(9)	8.8(12)	9.9(14)	10.5(16)	11.2(17)	$r_2 = 68$
3:	5.191)	5.7(2)	5.9(4)	6.6(8)	8.2(11)	$r_3 = 26$
4:	9.5(13)	1.0.3(15)	11.7(18)	12.1(19)	12.4(20)	$r_4 = 85$

H_o will be rejected at level .10 if $k \geq \chi^2_{.10,3} = 6.251$. The computed value of k is

$$k = \frac{12}{20(21)} \left[\frac{31^2 + 68^2 + 26^2 + 85^2}{5} \right] - 3(21) = 14.06 . \text{ Since}$$

$14.06 \geq 6.251$, reject H_o.

25. $H_0 : \mu_1 = \mu_2 = \mu_3$ will be rejected at level .05 if $k \geq \chi^2_{.05,2} = 5.992$. The ranks are 1, 3, 4, 5, 6, 7, 8, 9, 12, 14 for the first sample; 11, 13, 15, 16, 17, 18 for the second; 2, 10, 19, 20, 21, 22 for the third; so the rank totals are 69, 90, and 94.

$$k = \frac{12}{22(23)} \left[\frac{69^2}{10} + \frac{90^2}{6} + \frac{94^2}{5} \right] - 3(23) = 9.23 . \text{ Since } 9.23 \geq 5.992, \text{ we}$$

reject H_o.

27.

	1	2	3	4	5	6	7	8	9	10	r_i	r_i^2
I	1	2	3	3	2	1	1	3	1	2	19	361
H	2	1	1	2	1	2	2	1	2	3	17	289
C	3	3	2	1	3	3	3	2	3	1	24	576
												1226

The computed value of F_r is $\dfrac{12}{10(3)(4)}(1226) - 3(10)(4) = 2.60$, which is not

$\geq \chi^2_{.05,2} = 5.992$, so don't reject H_o.

Supplementary Exercises

29. Friedman's test is appropriate here. At level .05, H_o will be rejected if
$f_r \geq \chi^2_{.05,3} = 7.815$. It is easily verified that $r_{1.} = 28$, $r_{2.} = 29$, $r_{3.} = 16$,
$r_{4.} = 17$, from which the defining formula gives $f_r = 9.62$ and the computing
formula gives $f_r = 9.67$. Because $f_r \geq 7.815$, $H_0 : \alpha_1 = \alpha_2 = \alpha_3 = \alpha_4 = 0$
is rejected, and we conclude that there are effects due to different years.

31. From Table A.16, m = n = 5 implies that c = 22 for a confidence level of 95%, so
$mn - c + 1 = 25 - 22 = 1 = 4$. Thus the confidence interval extends from the 4th
smallest difference to the 4th largest difference. The 4 smallest differences are –7.1, -
6.5, -6.1, -5.9, and the 4 largest are –3.8, -3.7, -3.4, -3.2, so the C.I. is (-5.9, -3.8).

33.

    a. With "success" as defined, then Y is a binomial with n = 20. To determine the
        binomial proportion "p" we realize that since 25 is the hypothesized median, 50%
        of the distribution should be above 25, thus p = .50. From the Binomial Tables
        (Table A.1) with n = 20 and p = .50, we see that
$$\alpha = P(Y \geq 15) = 1 - P(Y \leq 14) = 1 - .979 = .021 .$$

    b. From the same binomial table as in **a**, we find that
$$P(Y \geq 14) = 1 - P(Y \leq 13) = 1 - .942 = .058 \text{ (a close as we can get to}$$
        .05),   so c = 14. For this data, we would reject H_o at level .058 if $Y \geq 14$. Y
        = (the number of observations in the sample that exceed 25) = 12, and since 12 is
        not ≥ 14, we fail to reject H_o.

35.

Sample:	y	x	y	y	x	x	x	y	y
Observations:	3.7	4.0	4.1	4.3	4.4	4.8	4.9	5.1	5.6
Rank:	1	3	5	7	9	8	6	4	2

The value of W' for this data is $w' = 3 + 6 + 8 + 9 = 26$. At level .05, the critical
value for the upper-tailed test is (Table A.14, m = 4, n = 5) c = 27 ($\alpha = .056$).
Since 26 is not ≥ 27, H_o cannot be rejected at level .05.

Chapter 15: Distribution-Free Procedures

CHAPTER 16

Section 16.1

1. All ten values of the quality statistic are between the two control limits, so no out-of-control signal is generated.

3. P(10 successive points inside the limits) = P(1st inside) x P(2nd inside) x...x P(10th inside) = $(.998)^{10}$ = .9802. P(25 successive points inside the limits) = $(.998)^{25}$ = .9512. $(.998)^{52}$ = .9011, but $(.998)^{53}$ = .8993, so for 53 successive points the probability that at least one will fall outside the control limits when the process is in control is 1 - .8993 = .1007 > .10.

Section 16.2

5.

a. P(point falls outside the limits when $\mu = \mu_0 + .5\sigma$)

$$= 1 - P\left(\mu_0 - \frac{3\sigma}{\sqrt{n}} < \overline{X} < \mu_0 + \frac{3\sigma}{\sqrt{n}} when\mu = \mu_0 + .5\sigma \right)$$
$$= 1 - P\left(-3 - .5\sqrt{n} < Z < 3 - .5\sqrt{n} \right)$$
$$= 1 - P\left(-4.12 < Z < 1.882 \right) = 1 - .9699 = .0301 .$$

b. $1 - P\left(\mu_0 - \frac{3\sigma}{\sqrt{n}} < \overline{X} < \mu_0 + \frac{3\sigma}{\sqrt{n}} when\mu = \mu_0 - \sigma \right)$

$$= 1 - P\left(-3 + \sqrt{n} < Z < 3 + \sqrt{n} \right) = 1 - P\left(-.76 < Z < 5.24 \right) = .2236$$

c. $1 - P\left(-3 - 2\sqrt{n} < Z < 3 - 2\sqrt{n} \right) = 1 - P\left(-7.47 < Z < -1.47 \right) = .6808$

7. $\overline{\overline{x}} = 12.95$ and $\overline{s} = .526$, so with $a_5 = .940$, the control limits are

$$12.95 \pm 3\frac{.526}{.940\sqrt{5}} = 12.95 \pm .75 = 12.20, 13.70 .$$ Again, every point $\left(\overline{x} \right)$ is

between these limits, so there is no evidence of an out-of-control process.

9. $\bar{\bar{x}} = \dfrac{2317.07}{24} = 96.54$, $\bar{s} = 1.264$, and $a_6 = .952$, giving the control limits

$96.54 \pm 3\dfrac{1.264}{.952\sqrt{6}} = 96.54 \pm 1.63 = 94.91, 98.17$. The value of $\bar{x}$ on the 22nd

day lies above the UCL, so the process appears to be out of control at that time.

11.

a. $P\left(\mu_0 - \dfrac{2.81\sigma}{\sqrt{n}} < \bar{X} < \mu_0 + \dfrac{2.81\sigma}{\sqrt{n}} \text{ when } \mu = \mu_0 \right)$

$= P(-2.81 < Z < 2.81) = .995$, so the probability that a point falls outside

the limits is .005 and $ARL = \dfrac{1}{.005} = 200$.

b. P = P(a point is outside the limits)

$= 1 - P\left(\mu_0 - \dfrac{2.81\sigma}{\sqrt{n}} < \bar{X} < \mu_0 + \dfrac{2.81\sigma}{\sqrt{n}} \text{ when } \mu = \mu_0 + \sigma \right)$

$= 1 - P\left(-2.81 - \sqrt{n} < Z < 2.81 - \sqrt{n} \right)$

$= 1 - P(-4.81 < Z < .81) = 1 - .791 = .209$. Thus $ARL = \dfrac{1}{.209} = 4.78$

c. $1 - .9974 = .0026$ so $ARL = \dfrac{1}{.0026} = 385$ for an in-control process, and when

$\mu = \mu_0 + \sigma$, the probability of an out-of-control point is

$1 - P(-3 - 2 < Z < 1) = 1 - P(Z < 1) = .1587$, so

$ARL = \dfrac{1}{.1587} = 6.30$.

13. $\bar{\bar{x}} = 12.95$, IQR = .4273, $k_s = .990$. The control limits are

$12.95 \pm 3\dfrac{.4273}{.990\sqrt{5}} = 12.45, 13.45 = 12.37, 13.53$.

Chapter 16: Quality Control Methods

Section 16.3

15.

    a. $\bar{r} = \dfrac{85.2}{30} = 2.84$, $b_4 = 2.058$, and $c_4 = .880$. Since n = 4, LCL = 0 and

$$\text{UCL} = 2.84 + \dfrac{3(.880)(2.84)}{2.058} = 2.84 + 3.64 = 6.48.$$

    b. $\bar{r} = 3.54$, $b_8 = 2.844$, and $c_8 = .820$, and the control limits are

$$= 3.54 \pm \dfrac{3(.820)(3.54)}{2.844} = 3.54 \pm 3.06 = .48, 6.60.$$

17. $\bar{s} = 1.2642$, $a_6 = .952$, and the control limits are

$$1.2642 \pm \dfrac{3(1.2642)\sqrt{1-(.952)^2}}{.952} = 1.2642 \pm 1.2194 = .045, 2.484.$$ The

smallest s_I is $s_{20} = .75$, and the largest is $s_{12} = 1.65$, so every value is between .045 and 2.434. The process appears to be in control with respect to variability.

Section 16.4

19. $\bar{p} = \Sigma \dfrac{\hat{p}_i}{k}$ where $\Sigma \hat{p}_i = \dfrac{x_1}{n} + \ldots + \dfrac{x_k}{n} = \dfrac{x_1 + \ldots + x_k}{n} = \dfrac{578}{100} = 5.78$. Thus

$$\bar{p} = \dfrac{5.78}{25} = .231.$$

    a. The control limits are $.231 \pm 3\sqrt{\dfrac{(.231)(.769)}{100}} = .231 \pm .126 = .105, .357.$

    b. $\dfrac{13}{100} = .130$, which is between the limits, but $\dfrac{39}{100} = .390$, which exceeds the

    upper control limit and therefore generates an out-of-control signal.

21. LCL > 0 when $\bar{p} > 3\sqrt{\dfrac{\bar{p}(1-\bar{p})}{n}}$, i.e. (after squaring both sides)

$50\bar{p}^2 > 3\bar{p}(1-\bar{p})$, i.e. $50\bar{p} > 3(1-\bar{p})$, i.e. $53\bar{p} > 3 \Rightarrow \bar{p} = \dfrac{3}{53} = .0566$.

23. $\Sigma x_i = 102$, $\bar{x} = 4.08$, and $\bar{x} \pm 3\sqrt{\bar{x}} = 4.08 \pm 6.06 \approx (-2.0, 10.1)$. Thus LCL = 0 and UCL = 10.1. Because no x_i exceeds 10.1, the process is judged to be in control.

25. With $u_i = \dfrac{x_i}{g_i}$, the u_i's are 3.75, 3.33, 3.75, 2.50, 5.00, 5.00, 12.50, 12.00, 6.67, 3.33, 1.67, 3.75, 6.25, 4.00, 6.00, 12.00, 3.75, 5.00, 8.33, and 1.67 for I = 1, ..., 20, giving $\bar{u} = 5.5125$. For $g_i = .6$, $\bar{u} \pm 3\sqrt{\dfrac{\bar{u}}{g_i}} = 5.5125 \pm 9.0933$, LCL = 0,

UCL = 14.6. For $g_i = .8$, $\bar{u} \pm 3\sqrt{\dfrac{\bar{u}}{g_i}} = 5.5125 \pm 7.857$, LCL = 0, UCL = 13.4.

For $g_i = 1.0$, $\bar{u} \pm 3\sqrt{\dfrac{\bar{u}}{g_i}} = 5.5125 \pm 7.0436$, LCL = 0, UCL = 12.6. Several u_i's are close to the corresponding UCL's but none exceed them, so the process is judged to be in control.

Section 16.5

27. $\mu_0 = 16$, $k = \dfrac{\Delta}{2} = 0.05$, $h = .20$, $d_i = \max(0, d_{i-1} + (\bar{x}_i - 16.05))$,

$e_i = \max(0, e_{i-1} + (\bar{x}_i - 15.95))$.

i	$\bar{x}_i - 16.05$	d_i	$\bar{x}_i - 15.95$	e_i
1	-0.058	0	0.024	0
2	0.001	0.001	0.101	0
3	0.016	0.017	0.116	0
4	-0.138	0	-0.038	0.038
5	-0.020	0	0.080	0
6	0.010	0.010	0.110	0
7	-0.068	0	0.032	0
8	-0.151	0	-0.054	0.054
9	-0.012	0	0.088	0
10	0.024	0.024	0.124	0
11	-0.021	0.003	0.079	0
12	-0.115	0	-0.015	0.015
13	-0.018	0	0.082	0
14	-0.090	0	0.010	0
15	0.005	0.005	0.105	0

For no time r is it the case that $d_r > .20$ or that $e_r > .20$, so no out-of-control signals are generated.

29. Connecting 600 on the in-control ARL scale to 4 on the out-of-control scale and extending to the k' scale gives k' = .87. Thus $k' = \dfrac{\Delta/2}{\sigma/\sqrt{n}} = \dfrac{.002}{.005/\sqrt{n}}$ from

which $\sqrt{n} = 2.175 \Rightarrow n = 4.73 = s$. Then connecting .87 on the k' scale to 600 on the out-of-control ARL scale and extending to h' gives h' = 2.8, so

$$h = \left(\dfrac{\sigma}{\sqrt{n}}\right)(2.8) = \left(\dfrac{.005}{\sqrt{5}}\right)(2.8) = .00626.$$

Section 16.6

31. For the binomial calculation, n = 50 and we wish

$$P(X \leq 2) = \binom{50}{0} p^0 (1-p)^{50} + \binom{50}{1} p^1 (1-p)^{49} + \binom{50}{2} p^2 (1-p)^{48}$$

$$= (1-p)^{50} + 50p(1-p)^{49} + 1225 p^2 (1-p)^{48} \text{ when p = .01, .02, ..., .10. For}$$

the hypergeometric calculation

$$P(X \leq 2) = \frac{\binom{M}{0}\binom{500-M}{50}}{\binom{500}{50}} + \frac{\binom{M}{1}\binom{500-M}{49}}{\binom{500}{50}} + \frac{\binom{M}{2}\binom{500-M}{48}}{\binom{500}{50}}, \text{ to be}$$

calculated for M = 5, 10, 15, ..., 50. The resulting probabilities appear in the answer section in the text.

33. $$P(X \leq 2) = \binom{100}{0} p^0 (1-p)^{100} + \binom{100}{1} p^1 (1-p)^{99} + \binom{100}{2} p^2 (1-p)^{98}$$

p	.01	.02	.03	.04	.05	.06	.07	.08	.09	.10
$P(X \leq 2)$	.9206	.6767	.4198	.2321	.1183	.0566	.0258	.0113	.0048	.0019

For values of p quite close to 0, the probability of lot acceptance using this plan is larger than that for the previous plan, whereas for larger p this plan is less likely to result in an "accept the lot" decision (the dividing point between "close to zero" and "larger p" is someplace between .01 and .02). In this sense, the current plan is better.

35. P(accepting the lot) = P(X_1 = 0 or 1) + P(X_1 = 2, X_2 = 0, 1, 2, or 3) + P(X_1 = 3, X_2 = 0, 1, or 2) = P(X_1 = 0 or 1) + P(X_1 = 2)P(X_2 = 0, 1, 2, or 3) + P(X_1 = 3)P(X_2 = 0, 1, or 2).

$$p = .01: = .9106 + (.0756)(.9984) + (.0122)(.9862) = .9981$$
$$p = .05: = .2794 + (.2611)(.7604) + (.2199)(.5405) = .5968$$
$$p = .10: = .0338 + (.0779)(.2503) + (.1386)(.1117) = .0688$$

37.

a. $AOQ = pP(A) = p[(1-p)^{50} + 50p(1-p)^{49} + 1225p^2(1-p)^{48}]$

p	.01	.02	.03	.04	.05	.06	.07	.08	.09	.10
AOQ	.010	.018	.024	.027	.027	.025	.022	.018	.014	.011

b. p = .0447, AOQL = .0447P(A) = .0274

c. ATI = 50P(A) + 2000(1 – P(A))

p	.01	.02	.03	.04	.05	.06	.07	.08	.09	.10
ATI	77.3	202.1	418.6	679.9	945.1	1188.8	1393.6	1559.3	1686.1	1781.6

Supplementary Exercises

39. $n = 6, k = 26, \Sigma \bar{x}_i = 10,980, \bar{\bar{x}} = 422.31, \Sigma s_i = 402, \bar{s} = 15.4615$,

$\Sigma r_i = 1074, \bar{r} = 41.3077$

S chart: $15.4615 \pm \dfrac{3(15.4615)\sqrt{1-(.952)^2}}{.952} = 15.4615 \pm 14.9141 \approx .55, 30.37$

R chart: $41.31 \pm \dfrac{3(.848)(41.31)}{2.536} = 41.31 \pm 41.44$, so LCL = 0, UCL = 82.75

$\bar{X}$ chart based on $\bar{s}$: $422.31 \pm \dfrac{3(15.4615)}{.952\sqrt{6}} = 402.42, 442.20$

$\bar{X}$ chart based on $\bar{r}$: $422.31 \pm \dfrac{3(41.3077)}{2.536\sqrt{6}} = 402.36, 442.26$

41.

i	$\bar{x}_i$	s_i	r_i
1	50.83	1.172	2.2
2	50.10	.854	1.7
3	50.30	1.136	2.1
4	50.23	1.097	2.1
5	50.33	.666	1.3
6	51.20	.854	1.7
7	50.17	.416	.8
8	50.70	.964	1.8
9	49.93	1.159	2.1
10	49.97	.473	.9
11	50.13	.698	.9
12	49.33	.833	1.6
13	50.23	.839	1.5
14	50.33	.404	.8
15	49.30	.265	.5
16	49.90	.854	1.7
17	50.40	.781	1.4
18	49.37	.902	1.8
19	49.87	.643	1.2
20	50.00	.794	1.5
21	50.80	2.931	5.6
22	50.43	.971	1.9

$\Sigma s_i = 19.706$, $\bar{s} = .8957$, $\Sigma \bar{x}_i = 1103.85$, $\bar{\bar{x}} = 50.175$, $a_3 = .886$, from which an s chart has LCL = 0 and UCL =

$.8957 + \dfrac{3(.8957)\sqrt{1-(.886)^2}}{.886} = 2.3020$, and $s_{21} = 2.931 > UCL$. Since an

assignable cause is assumed to have been identified we eliminate the 21st group.
Then $\Sigma s_i = 16.775$, $\bar{s} = .7998$, $\bar{\bar{x}} = 50.145$. The resulting UCL for an s chart
is 2.0529, and $s_i < 2.0529$ for every remaining i. The $\bar{x}$ chart based on $\bar{s}$ has

limits $50.145 \pm \dfrac{3(.7988)}{.886\sqrt{3}} = 48.58, 51.71$. All $\bar{x}_i$ values are between these limits.

43. $\Sigma n_i = 4(16) + (3)(4) = 76$, $\Sigma n_i \bar{x}_i = 32,729.4$, $\bar{\bar{x}} = 430.65$,

$s^2 = \dfrac{\Sigma(n_i - 1)s_i^2}{\Sigma(n_i - 1)} = \dfrac{27,380.16 - 5661.4}{76 - 20} = 590.0279$, so s = 24.2905. For

variation: when n = 3,

$UCL = 24.2905 + \dfrac{3(24.2905)\sqrt{1 - (.886)^2}}{.886} = 24.29 + 38.14 = 62.43$, when

n = 4, $UCL = 24.2905 + \dfrac{3(24.2905)\sqrt{1 - (.921)^2}}{.921} = 24.29 + 30.82 = 55.11$.

For location: when n = 3, $430.65 \pm 47.49 = 383.16, 478.14$, and when n = 4,

$430.65 \pm 39.56 = 391.09, 470.21$.